中外建筑英语选读

主　编　丁　虹　王彬彬

副主编　何飞云　张　红　赵佳娜　李桂平

電子工業出版社
Publishing House of Electronics Industry
北京 · BEIJING

内容简介

本书以英语为载体，内容包含中国古建筑、“一带一路”沿线地标建筑、现代建筑、知名建筑师、建筑企业、建筑奖项六个模块。

本书可作为高职高专建筑专业英语课程的专用教材，也可作为各级各类学校的选修课教材和学生课外阅读材料，同时也适合对建筑文化英语感兴趣的社会各界人士使用。

图书在版编目（CIP）数据

中外建筑英语选读 / 丁虹，王彬彬主编 .—北京：电子工业出版社，2023.2
ISBN 978-7-121-45013-6

Ⅰ.①中… Ⅱ.①丁… ②王… Ⅲ.①建筑—英语 Ⅳ.① TU

中国国家版本馆 CIP 数据核字（2023）第 016817 号

责任编辑：王　花
印　　刷：天津画中画印刷有限公司
装　　订：天津画中画印刷有限公司
出版发行：电子工业出版社
　　　　　北京市海淀区万寿路 173 信箱　邮编 100036
开　　本：787 × 1092　1/16　印张：7.5　字数：249.6 千字
版　　次：2023 年 2 月第 1 版
印　　次：2023 年 2 月第 1 次印刷
定　　价：36.00 元

凡所购买电子工业出版社图书有缺损问题，请向购买书店调换。若书店售缺，请与本社发行部联系，联系及邮购电话：（010）88254888，85258888。

质量投诉请发邮件至 zlts@phei.com.cn，盗版侵权举报请发邮件至 dbqq@phei.com.cn。

本书咨询联系方式：（010）88254609 或 hzh@phei.com.cn。

前言

英语课程作为高等职业教育专科课程体系的有机组成部分，兼具工具性与人文性，旨在培养学生学习英语和应用英语的能力并落实立德树人根本任务，通过课程学习使学生能够识别、理解、尊重世界多元文化，拓宽国际视野，坚定文化自信，树立中华民族共同体意识和人类命运共同体意识。为此，本书编写组以最新的《高等职业教育专科英语课程标准》为指导方针，对高职建筑类的在校生、毕业生和用人单位进行了调研，在了解学生需求和社会需求的基础上结合学科的核心素养要求确定了内容模块和思考练习，帮助学生在具有基本的听说能力的基础上，提高阅读和翻译英文资料的能力，并为未来继续学习和终身发展奠定良好的英语基础。

《中外建筑英语选读》侧重于提高书本使用者的建筑行业综合素养（如建筑人文知识、思想、方法、精神等）和英语水平，教学模块既有建筑行业相关的基础知识，也有建筑领域先贤今哲的实践案例。本书涵盖了建筑行业古今中外不同的素材，包含了建筑领域的多个环节，丰富多样，适合高职高专的学生学习，也适合对建筑英语感兴趣的社会各界人士学习使用。

本书以“活页读本”的方式呈现，体现以下特色。

1．前瞻：适时动态更新调整，“活”化教学内容。

2．革新：选取短篇经典文章，符合当下的碎片化学习习惯。

3．修德：厚植家国情怀，实现全过程育人。

4．丰富：线上线下融合，打造立体化资源。

5．巩固：练习设计既提升文化内涵，又培养辩证思考能力。

本书由丁虹、王彬彬主编，何飞云、张红、赵佳娜、李桂平共同编写了中国古建筑、“一带一路”沿线地标建筑、现代建筑、知名建筑师、建筑企业、建筑奖项六个模块。

特别感谢浙江建设职业技术学院袁荣儿教授及浙江工商大学王镇博士对本书提出的宝贵意见。

由于编者水平有限，疏漏和不妥之处在所难免，恳请读者不吝指正。

编者

2022 年 10 月

目 录

Chapter 1

Ancient Chinese Architecture

Preview:

Ancient buildings not only reflect the glory of national history, the great achievements of its culture, but also reflect the art, science and technology in the past. Ancient Chinese architecture experiences a long history, leading to its unique aesthetics and characteristics. When we look at these ancient masterpieces again, we are still amazed by their beauty and complexity, and astonished by the intelligence of our ancestors. All of these will arouse our patriotism and national pride. In this chapter, we are going to appreciate three famous ancient Chinese architecture.

Text A

The Great Wall

The Great Wall, the largest man-made project in the world, is one of the most magnificent ancient military **fortifications** in northern China, stretching almost 21,000 kilometers in total from west to east of China.

Although called the wall, it is actually an integrated defense system, consisting of solid walls, watchtowers, **shelters** on the wall, passes along the walls and so on. Most of the wall is about 7.8 meters tall and 5.8 meters in width at the top.

Dating back to 211BC, Qin Shi Huang, the first emperor of China, ordered to build the Great Wall by connecting and extending several ancient states' boundary walls. It was constantly consolidated and extended by many dynasties since then.

The Great Wall was inscribed on the UNESCO World Heritage List in 1987.With a history of more than 2,000 years, some of the sections are now in ruins. However, the Great Wall is still one of the most appealing attractions all around the world owing to its grand **architecture** and glorious history.

New Words

单词	音标	词义
magnificent	/mæg'nɪfɪsnt/	*adj.* 壮丽的，伟大的，华丽的
stretch	/stretʃ/	*vt.* 伸展，张开；*vi.* 伸展 *adj.* 可伸缩的；*n.* 伸展，延伸

续表

单词	音标	词义
integrated	/'ɪntɪgreɪtɪd /	*adj.* 整体的；综合的；连成一体的
extend	/ɪk'stend/	*v.* [I or T] (空间、时间等) 延伸 *vt.* 延长；扩展
boundary	/'baʊnd(ə)rɪ/	*n.* [C] 分界线；界线，范围
consolidate	/kən'sɒlɪdeɪt/	*v.* [I or T] (使) 巩固；(使) 加强
appealing	/ə'pi:lɪŋ/	*adj.* 有魅力的；动人的；吸引人的；上诉的；哀求的
glorious	/'glɔ:rɪəs/	*adj.* 荣誉的，光荣的；辉煌的

Phrases

词组	词义
date back to	追溯到；从……开始有
in ruin	成为废墟；严重受损的
owe to	归功于

Background Information

Watchtower

烽火台，也被称为瞭望塔或警戒塔，建在中国长城上或沿长城修建，用来监视敌人和传递军事信息。在古代，如果有入侵者靠近，城楼上的士兵会在白天制造烟雾、在晚上生火来警示自己的军队。烽火台是长城防御系统的重要组成部分，它们是中国最古老但高效的“电报系统”。

UNESCO World Heritage List

联合国教科文组织世界遗产名录旨在鼓励识别、保护和保存世界各地独特的、不可替代的自然和文化遗产。截至2022年，中国共有56处自然和文化遗产被列入联合国教科文组织世界遗产名录。

Architectural Terms

1. **fortification** /'fɔːtɪfɪ'keɪʃ(ə)n/ *n.* [C usually plural] strong walls, towers, etc. that are built to protect a place 防御工事
   - 例句：Some of the old *fortifications* still exist.
     一些古老的防御工事仍然存在。

2. **shelter** /'ʃeltə/
   - ➢ *n.* [C] a building where people or animals that have nowhere to live or that are in danger can stay and receive help 避难所
     - 例句：We use bricks and branches of trees to form a simple *shelter*.
       我们用砖和树枝搭成一个简陋的避难所。
   - ➢ *n.* [U] a place to live, considered as one of the basic human needs 住所
     - 例句：Food, clothing, *shelter* and transportation are essential to life.
       衣、食、住、行是生活中必不可少的。

3. **architecture** /'ɑːkɪtektʃə(r)/
   - ➢ *n.* [U] the art and practice of designing and making buildings 建筑学；建筑艺术
     - 例句：He majored in *architecture* at university.
       他在大学里主修建筑学。
   - ➢ *n.* [U] the style in which buildings are made 建筑风格
     - 例句：This is a wonderful example of Chinese-style *architecture*.
       这是中式建筑的一个极好的例子。
     - 近义词：
     - ✧ construction *n.* [U] the way that something has been built or made 强调建设或构造
     - ✧ structure *n.* [U] the way in which the parts of something are connected together or arranged 多指结构
     - ✧ building *n.* [C] a structure such as house, office or factory that has a roof and four walls（具体的）建筑物；房屋；楼房

Difficult Sentences

1. The Great Wall, the largest man-made project in the world, is one of the most magnificent ancient military fortifications in northern China, stretching almost 21,000 kilometers in total from west to east of China.

Sentence analysis

1）*the largest man-made project in the world* 是插入语，表示长城是世界上最大型的人造工程。

2）*one of the most magnificent* 是最高级的用法之一，表示“最宏伟的……之一”。

3）现在分词 *stretching* 作伴随状语，补充说明了长城之长。

Translation

长城是世界上最大型的人造工程，是中国北方最宏伟的古代军事防御工事之一，从中国的西部到东部全长近 21000 公里。

2. Although called the wall, it is actually an integrated defense system, consisting of solid walls, watchtowers, shelters on the wall, passes along the walls and so on.

Sentence analysis

1）此句是由 *although* 引导的让步状语从句。

2）*called* 为过去分词，表被动。

3）现在分词 *consisting of* 作伴随状语，表示这个系统的组成部分。

Translation

长城虽然被称为城墙，但其实是一个完整的防御体系，由坚实的城墙、烽火台、城墙上的掩体、沿城墙的通道等构成。

Text B

Yingxian Wooden Pagoda

Built in the Liao Dynasty, Yingxian Wooden **Pagoda** in Ying County, also called the Sakyamuni Pagoda of Fogong Temple, is the existing oldest timber multi-storied tower in China. It is even known as one of Three Wonders of Towers and declared as the world's highest wooden pagoda by Guinness World Records in 2016.

Standing 67.31 meters tall with a **base** diameter of 30.27 meters, the tower weighs about 2,600 tons with 3,000 cubic meters of wood. As a masterpiece of Chinese wooden construction, this beautiful octagonal-shaped tower was put together without using a single nail.

From the exterior, the pagoda seems to have only five stories, yet its interior shows that it has nine stories in all, with four hidden stories under **eaves** to prevent displacement. The pagoda employs 54 different kinds of dougong (**bracket** supports) in **construction**, which is one of the most complex examples of the roof structure in China, indeed, in the world.

According to researches, the pagoda has withstood numerous earthquakes (a dozen times with magnitude over 5) through the centuries.

Yingxian Wooden Pagoda

New Words

单词	音标	词义
timber	/'tɪmbə/	*n.* [C or U] 木料，木材
declare	/dɪ'kleə/	*vt.* 宣布，宣告；声明 *vi.* 申报

续表

单词	音标	词义
diameter	/daɪ'æmɪtə/	*n.* [C] 直径；放大率
octagonal	/ɔk'tægənəl/	*adj.* 八边形的
exterior	/ɪk'stɪərɪə/	*adj.* 外部的；表面的；外在的 *n.* [C] 外部；表面；外貌
interior	/ɪn'tɪɜ:rɪə/	*n.* [C] 内部；内地 *adj.* 内部的；内地的，国内的
displacement	/dɪs'pleɪsm(ə)nt/	*n.* [U] 移置，代替
withstand	/wɪð'stænd/	*vt.* 经受，承受
magnitude	/'mægnɪtju:d/	*n.* [U] 巨大；重要性 *n.* [C] 震级

Phrases

词组	词义
cubic meter	立方米

Background Information

Three Wonders of Towers

世界三大“奇迹塔”分别是中国的应县木塔、意大利的比萨斜塔和法国的埃菲尔铁塔。

Yingxian Wooden Pagoda

Leaning Tower of Pisa

Eiffel Tower

Guinness World Records

《吉尼斯世界纪录大全》是一本每年再版一次的参考书，罗列了人类成就和自然界极端情况的世界纪录。这本书是休·比弗爵士的创意，1954 年 8 月，他与双胞胎兄弟诺里斯·麦克沃特和罗斯·麦克沃特在伦敦舰队街共同编写了这本书。

Dougong

斗拱，托架臂，也被称为枓栱，是插在柱子顶部和横梁之间的支架系统。斗拱是中国古代木结构建筑一种特有的构件。

Architectural Terms

1. **pagoda** /pə'gəʊdə/ *n.* [C] a tall religious building in Asia with many levels, each of which has a curved roof（亚洲的）塔，宝塔
   - 例句：We could see the sun setting behind the white *pagoda*.
     我们能亲眼看到太阳在白塔后西沉。
   - 近义词：

   pagoda 和 tower 都包含或表示“塔”的意思，但两者有区别。

   ✧ pagoda 多指(佛教、印度教的多层)宝塔、浮屠，塔的每层都有屋顶。

   ✧ tower 高的、窄的建筑都可以称作塔或者高楼，可作为单个建筑物或某个建筑物（如城堡等）的一部分。

2. **base** /beɪs/ *n.* [C] the bottom part of an object, on which it rests, or the lowest part of something 基底，底座；底层，底子
   - 例句：There is a door at the *base* of the tower.
     这座塔的底座有一扇门。
   - 近义词：

     basis /'beɪsɪs/ *n.* [C] the most important facts, ideas, etc. from which something is developed 基础；根据

3. **eave** /iːv/ *n.* [C usually plural] the edge of a roof that sticks out over the top of a wall 屋檐
   - 例句：The overhanging *eaves* are supported by six square wooden pillars.
     悬挑屋檐由六根方形木柱支撑。
   - 近义词：

     roof /ruːf/ *n.* [C] the covering that forms the top of a building, vehicle, etc. 应用范围较广，不止屋顶，还可以表示车顶、顶部、盖子等。

4. **bracket** /'brækɪt/ *n.* [C] a piece of metal, wood, or plastic, often in the shape of the letter L, fixed to a wall to support something such as a shelf（通常呈 L 形、固定在墙上的）托架；支架；小壁架
   - 例句：Each *bracket* is fixed to the wall with just three screws.
     每个托架只用 3 颗螺钉固定在墙上。

5. **construction** /kən'strʌkʃn/ *n.* [U] the particular type of structure, materials, etc. that something has（某种）构造；结构
   - 例句：The bridge is of lightweight *construction*.
     这座桥结构较轻。

Polysemous Words

单词	常用词义	本文词义
story	*n.* [C] 故事	*n.* [C] 楼层
stand	*vi.* 站立	*link.* 高度为，高达

Difficult Sentences

Standing 67.31 meters tall with a base diameter of 30.27 meters, the tower weighs about 2,600 tons with 3,000 cubic meters of wood.

Sentence analysis

此句是由现在分词 *standing* 引导的状语从句，描述古塔的一些基本数据，包括塔高、塔的地基、用料和重量等。

Translation

释迦塔塔高 67.31 米，底层直径 30.27 米，全塔耗材红松木料 3000 立方米，重达 2600 多吨。

Text C

Zhaozhou Bridge

Located in Zhao County, Hebei Province, Zhaozhou Bridge (also known as Anji Bridge and Great Stone Bridge) is the oldest and best-preserved open-spandrel stone segmental **arch** bridge in the world. It is constructed around 605 of Sui Dynasty (581-618) by famous architect Li Chun, with more than 1,400 years of history.

The bridge stands 7.23 meters tall, 50.82 meters in length, 37.37 meters in **span**, and has a width of 9 meters. The main arch is segmented to make the slope of the bridge floor gentle and to help the passage of vehicles and horses. Meanwhile, the bridge bears two pairs of small circular arches at each side of the central arch to reduce the load of the **deck** down to the main arch, save the materials, and minimize the water blockage when Xiao River was flooded. Zhaozhou Bridge is the earliest bridge to use this unique design, 700 years earlier than its European counterparts.

Zhaozhou Bridge

To date, Zhaozhou Bridge has survived lots of geological disasters, which proves the delicacy and greatness of its design and construction.

New Words

单词	音标	词义
county	/'kaʊnti/	*n.* [C] 郡，县
preserve	/prɪ'zɜːv/	*vt.* 保存；保护；维持；禁猎；腌 *n.* [C or U] 果酱；禁猎地；保护区
spandrel	/'spændrɪl/	*n.* [C] 拱肩

续表

单词	音标	词义
segmental	/seg'ment(ə)l/	*adj.* 部分的
slope	/sləʊp/	*n.* [C] 倾斜；斜率；斜坡 *vi.* 倾斜 *vt.* 扛；倾斜；使倾斜
circular	/'sɜ:kjʊlə/	*adj.* 圆形的，环形的；*n.* [C] 通知，通告
load	/ləʊd/	*n.* [C] 负载，负荷 *v.* [I or T] 装，装载
counterpart	/'kaʊntəpɑ:t/	*n.* [C] 与对方地位相当的人，与另一方作用相当的物
geological	/dʒɪə'lɔdʒɪkəl/	*adj.* 地质（学）的
delicacy	/'delɪkəsɪ/	*n.* [C] 美食 *n.* [U] 精致，精美；易碎

Phrases

词组	词义
water blockage	水阻

Background Information

Li Chun

李春，隋代造桥匠师，现今河北邢台临城人士，是中国乃至世界建筑史上第一位桥梁专家。赵州桥凝聚了李春的汗水和心血。

Architectural Terms

1. **arch** /ɑ:tʃ/ *n.* [C] a structure, consisting of a curved top on two supports, that holds the weight of something above it 拱；拱形结构；拱门（或窗、顶、洞等）
   - 例句：Passing through the *arch*, you enter the courtyard.

穿过拱门，你就到了天井中。

- 近义词：

curve /kɜːv/ *n.* [C] a line that bends continuously and has no straight parts 曲线，弧线；转弯

2．**span** /spæn/ *n.* [C] the area of a bridge, etc. between two supports（桥等的）跨度，跨距，墩距

- 例句：The bridge crosses the river in a single *span*.

  河上的桥是单孔桥。

- 近义词：

distance /'dɪst(ə)ns/ *n.* [C or U] the amount of space between two places or things 两物或两个地方间的距离

3．**deck** /dek/ *n.* [C] one of the levels on a ship, bus, etc.（船或公共汽车等的）一层，层面

- 例句：But at present there are few special waterproofing materials for concrete bridge *decks* in China, which can't meet the demand.

  但中国目前桥面防水材料严重匮乏，不能满足需求。

- 近义词：

✧ roadway /'rəʊdweɪ/ *n.* [C] the part of the road on which vehicles drive 车行道

✧ platform /'plætfɔːm/ *n.* [C] a flat raised area or structure 平台

这三个词虽然都表示某一个平面，但侧重点不同。roadway 一般专指车道；platform 可指突起的一个大平台，即可指月台或工作平台等；deck 除了表示桥面外，还可以表示船的甲板或者上下层。

Polysemous Words

单词	常用词义	本文词义
gentle	*adj.* 和蔼的；文雅的	*adj.* 平和的；徐缓的
passage	*n.* [C] 文章	*n.* [U] 通过，穿过
bear	*n.* [C] 熊	*vt.* 具有，带有

Difficult Sentences

1. Located in Zhao County, Hebei Province, Zhaozhou Bridge (also known as Anji Bridge and Great Stone Bridge) is the oldest and best-preserved open-spandrel stone segmental arch bridge in the world.

   Sentence analysis

   此句概述了赵州桥的特点。*oldest* 和 *best-preserved* 都采用了形容词最高级结构，*oldest* 意为“最古老的”，*best-preserved* 意为“保存最为完好的”。

   Translation

   赵州桥坐落于中国河北省赵县，又名安济桥、大石桥，是目前世界上最古老、保存最为完好的大跨度单孔坦弧敞肩石拱桥。

2. Meanwhile, the bridge bears two pairs of small circular arches at each side of the central arch to reduce the load of the deck down to the main arch, save the materials, and minimize the water blockage when Xiao River was flooded.

   Sentence analysis

   此句用了 3 个 *to do* 非谓语结构来说明赵州桥敞肩拱设计的作用。

   1) *reduce the load* 减轻负担

   2) *save the materials* 节省材料

   3) *minimize the water blockage* 减少水阻

   Translation

   同时，两侧拱肩部分各建两个对称的小拱，伏在主拱的肩上。这种“敞肩拱”结构形式可以减轻桥面对主拱的负担，节省石材，当洨河洪水来临时，还能减少水阻。

Quiz for Chapter 1

Major: ______________ No.: ______________ Name: ______________

1 VOCABULARY AND GRAMMAR

A Check the best meaning for the underlined word in each sentence.

1) The report examines teaching methods ***employed*** in the classroom.
   A. paid B. used C. spent D. hired

2) We want to ***preserve*** the character of the town while improving the facilities.
   A. prevent B. reserve C. keep and save D. observe

3) The stone plaque ***bearing*** his name was smashed to pieces.
   A. giving birth to B. an animal C. accepting D. having

4) The government prohibits the ***passage*** of foreign troops and planes across its territory.
   A. movement B. corridor C. journey D. part

B Choose the right word.

5) Although ____________ the wall, it is actually an integrated defense system, consisting of solid walls, watchtowers, shelters on the wall, passes along the walls and so on.
   A. to call B. called C. calling D. called

6) The middle section is the gem of the whole garden, one-third ____________ is covered by water.
   A. where B. which C. of which D. that

7) He carried out his duties with great ____________ and understanding.
   A. delicate B. delicacy C. delicious D. deliberation

8) The disease ____________ into the remote town.
   A. extend B. tend C. intend D. tent

9) This bridge is ____________ than that one.
   A. more longer B. many more longer
   C. not longer D. much more longer

2 COMMON KNOWLEDGE

10) Which is the longest wall in the world?

A. Hadrian's Wall. B. The Great Wall.

C. Wailing Wall. D. Berlin Wall.

11) Which does not belong to the Three Wonders of Towers?

A. the Tower of London. B. the Leaning Tower of Pisa in Italia.

C. the Eiffel Tower in France. D. the Yingxian Wooden Pagoda in China.

12) Which is the existing oldest timber pagoda in China?

A. Yingxian Wooden Pagoda. B. Big Wild Goose Pagoda.

C. Leifeng Pagoda. D. The White Pagoda in Beihai Park.

13) Who was the first Bridge architect in China?

A. Luban. B. Zou Wenkai. C. Li Jie. D. Li Chun.

14) Which is the existing oldest bridge in China?

A. Lugou Bridge. B. Zhaozhou Bridge.

C. Guangji Bridge. D. Luoyang Bridge.

15) Guinness World Records is ____________.

A. a book B. a magazine C. a TV programme D. a movie

16) Which architecture is not belonging to UNESCO World Heritage List?

A. Lushan National Park. B. Old Town of Lijiang.

C. Zhaozhou Bridge. D. West Lake Cultural Landscape of Hangzhou.

17) How many sites has been listed on the UNESCO World Heritage List in China?

A. 30. B. 40. C. 50. D. 56.

18) Which bridge is the longest one in the world?

A. Golden Gate Bridge. B. Hong Kong-Zhuhai-Macao Bridge.

C. Tower Bridge. D. Sydney Harbour Bridge.

Word Power

Major: ______________ No.: ______________ Name: ______________

Write the words and phrases under the pictures.

shelter	construction	arch
dougong	watchtower	timber

1.____________________

2.____________________

3.____________________

4.____________________

5.____________________

6.____________________

Group Work

Scan the webpage and complete the chart.

Try to find more information about ancient Chinese architecture.

Architecture	Useful Information
Potala Palace	Tips: 1. Location: ____________ 2. Construction: ____________ 3. Time: ____________ 4. Function: ____________
Mogao Grottoes	Tips: 1. Location: ____________ 2. Construction: ____________ 3. Time: ____________ 4. Function: ____________
Fujian Tulou	Tips: 1. Location: ____________ 2. Construction: ____________ 3. Time: ____________ 4. Function: ____________
____________ **Architecture**	Tips: 1. Location: ____________ 2. Construction: ____________ 3. Time: ____________ 4. Function: ____________

Critical Thinking

Is it necessary to protect the ancient Chinese architecture?

For reference:

Heritage is our legacy from the past, what we live with today, and what we pass on to future generations. Our cultural and natural heritage are both irreplaceable sources of life and inspiration. What makes the concept of World Heritage exceptional is its universal application. World Heritage sites belong to all the peoples of the world, irrespective of the territory on which they are located.

我们从先祖那里继承了世界遗产，以之为伴并将世代传承。世界文化遗产象征着蓬勃的生命力和不可替代的灵感源泉，使之与众不同的是它的普世价值。世界遗产属于世界各个民族，不论其在哪一片土地。

—— UNESCO

Architecture should speak of its time and place, but yearn for timelessness.

建筑应表达它所处的时间及地点，但向往永恒。

—— Frank Gehry

Appreciation

Archaeological Ruins of Liangzhu City
良渚古城遗址
in the List 2019

Beijing-Hangzhou Grand Canal
京杭大运河
in the List 2014

Lingyin Temple
灵隐寺
built in 326AD

Leifeng Pagoda
雷峰塔
built in 977AD

Self Assessment for Chapter 1

I can…

Very well OK A little

□ □ □ Use the architectural words in this part

□ □ □ Master some information of ancient Chinese architecture

□ □ □ Say something about ancient Chinese architecture

□ □ □ Recognize some sites in the UNESCO World Heritage List

Chapter 2

Architecture Along the Belt and Road

Preview:

The Belt and Road Initiative, a global infrastructure development strategy proposed by the Chinese government in 2013, has become a grand international cooperation project. The Belt and Road Initiative links nearly 70 countries and international organizations, among which there are a great number of landmark buildings, such as the Imperial Palace, Red Square and Egyptian Pyramids. These landmark buildings are expressive in styles, history, culture, art and other fields, and, of course, have become local representative buildings. Now, let's explore the mysteries of these landmark buildings!

Text A

The Imperial Palace

Situated in the heart of Beijing, the Imperial Palace, also called the Forbidden City, was built from 1406 to 1420 by the third emperor of the Ming Dynasty, the Yongle Emperor. During over five hundred years of imperial operation, the palace served as the **residence** and court of twenty-four emperors.

The Forbidden City is surrounded by 10-meter-high walls and a 52-meter-wide moat. Measuring 961 meters from north to south and 753 meters from east to west, the **complex** covers an area of 1,120,000 square meters.

Known as the Outer Court, the southern portion of the Forbidden City features three main halls. The Outer Court was the venue for the emperor's court and grand audiences. Mirroring this arrangement is the Inner Court, which is not only comprised of the residences of the emperor and his consorts but also venues for religious rituals and administrative activities.

These structures were designed in strict accordance to the traditional code of architectural hierarchy, which designated specific features to reflect the authority and status of the emperor.

New Words

单词	音标	词义
imperial	/ɪm'pɪəriəl/	*adj.* 帝国的，皇帝的

续表

单词	音标	词义
emperor	/'empərə(r)/	*n.* [C] 皇帝
portion	/'pɔːʃ(ə)n/	*n.* [C] 一部分，一份；（食物的）一份
venue	/'venjuː/	*n.* [C]（公共事件的）发生场所，举行地点；会场
consort	/'kɒnsɔːt/	*n.* [C]（尤指统治者的）配偶 *vi.*（尤指与品行不好的人）勾结；厮混
religious	/rɪ'lɪdʒəs/	*adj.* 宗教的，宗教上的；笃信宗教的
ritual	/'rɪtʃuəl/	*n.* [C or U]（尤指）仪式；例行公事，老规矩
administrative	/əd'mɪnɪstrətɪv/	*adj.* 管理的，行政的
hierarchy	/'haɪərɑːki/	*n.* [C] 等级制度；统治集团
designate	/'dezɪgneɪt/	*vt.* 指定，选定，委派；指定，划定（特征、用途）
reflect	/rɪ'flekt/	*vt.* 显示，反映，表达 *v.* [I or T] 反射（光、热、声等），反映，映出（影像）
authority	/ɔː'θɒrəti/	*n.* [U] 威信，权力；管辖权 *n.* [C] 当局，官方；权威人士，专家
status	/'steɪtəs/	*n.* [C or U]（尤指在社会中的）地位，身份

Phrases

词组	词义
serve as	担任……，充当……；起……的作用
be comprised of	由……组成
in accordance to	根据

Background Information

Yongle Emperor

永乐皇帝一般指明成祖朱棣，是明朝第三位皇帝，明太祖朱元璋第四子。永乐皇帝于建文四年（1402 年）即位，在位二十二年，年号“永乐”，统治期间，经济繁荣，国力强盛，史称“永乐盛世”。

Architectural Terms

1. **situate** /'sɪtʃueɪt/ *vt.* to put something in a particular position 使坐落于
   - 例句：They plan to *situate* the bus stop at the corner of the road.
     他们打算把公交车站设在道路拐角上。
   - 近义词：
     locate /ləʊ'keɪt/ *vt.* to be in a particular place 位于……；在……附近
     常见用法：be located in

2. **residence** /'rezɪdəns/ *n.* [C] a home 住所；住房；宅第
   - 例句：10 Downing Street is the British Prime Minister's official *residence*.
     唐宁街 10 号是英国首相的官邸。
   - 近义词：
     accommodation /əˌkɒmə'deɪʃ(ə)n/ *n.* [U] a place to live, work, stay, etc. in 住处；工作场所；停留处

3. **complex** /'kɒmpleks/ *n.* [C] a large building with various connected rooms or a related group of buildings 建筑群
   - 例句：There is an industrial *complex*.
     那里有一个工业建筑群。
   - 近义词：
     composite /'kɒmpəzɪt/ *n.* [C] something that is made of various different parts（由不同部分组成的）混合物；合成物；综合体

Polysemous Words

单词	常用词义	本文词义
court	*n.* [C] 法院；球场	*n.* [C or U] 宫廷
measure	*n.* [C usually plural] 方法，措施	*vt.* 测量
feature	*n.* [C] 特色，特征，特点	*vt.* 以……为特色；是……的特征

Difficult Sentences

1. Situated in the heart of Beijing, the Imperial Palace, also called the Forbidden City, was built from 1406 to 1420 by the third emperor of the Ming Dynasty, the Yongle Emperor.

Sentence analysis

1) *situated in the heart of Beijing* 是过去分词作状语，表示故宫所处的位置。

2) *also called the Forbidden City* 是 *the Imperial Palace* 的同位语，作用是补充说明故宫的另外一个名称。

3) 整个句子的主句是一个被动句，句子主干为 *The Imperial Palace was built*。

Translation

故宫又称紫禁城，位于北京市中心，由明朝第三位皇帝永乐皇帝于 1406 年至 1420 年修建。

2. These structures were designed in strict accordance to the traditional code of architectural hierarchy, which designated specific features to reflect the authority and status of the emperor.

Sentence analysis

which designated specific features to reflect the authority and status of the emperor 是非限制性定语从句，起修饰作用。限制性定语从句不可缺少先行词，去掉它主句意思往往不明确；非限制性定语从句是对先行词的附加说明，去掉它也不会影响主句的意思，它与主句之间通常用逗号分开。

Translation

这些建筑严格按照传统的建筑等级规范设计，这种建筑等级规范以某些特定的特征来反映皇帝的权威和地位。

Text B

Red Square

Red **Square**, the largest and most famous square in Russia, is the central square of Moscow and the symbolic center of all Russia. All the major streets of Moscow radiate from here.

Standing in Red Square, you can see the most significant buildings in the capital: the Kremlin, GUM department store, the State Historical Museum, Lenin's **Mausoleum**, and of course, St Basil's Cathedral.

Red Square has always been the main square of the city, and its history dates back to the 1490s when a new red-brick Kremlin was built in Moscow. The east side of the fortification was vulnerable since it was not protected by the rivers nor any other natural barriers. This area outside walls was cleared in order to create a field of fire for shooting and became Red Square.

Not many people realize that Red Square's name has nothing to do with Communism or Soviet Russia. The old Russian word for "beautiful" and "red" was the same, So "Red Square" means "Beautiful Square".

New Words

单词	音标	词义
symbolic	/sɪm'bɒlɪk/	*adj.* 代表的，象征的；（行为举动）象征性的
significant	/sɪg'nɪfɪkənt/	*adj.* 重要的，显著的

续表

单词	音标	词义
capital	/'kæpɪt(ə)l/	*n.* [C] 首都；大写字母 *adj.* 大写字母的
vulnerable	/'vʌlnərəbl/	*adj.* 易受伤的；易受影响（或攻击）的，脆弱的
barrier	/'bæriə(r)/	*n.* [C] 屏障，障碍物；阻碍，隔阂
communism	/'kɒmjunɪzəm/	*n.* [U] 共产主义

Background Information

Kremlin

克里姆林宫位于莫斯科的心脏地带，是俄罗斯联邦的象征，也是总统府所在地。克里姆林宫曾是俄罗斯沙皇的居所之一，享有“世界第八奇景”的美誉。

State Historical Museum

俄罗斯国家历史博物馆馆藏的文物不仅有超过 50 万年的旧石器时代物品，还包括反映现代俄罗斯历史变迁的重要展览物。该博物馆兴建于 19 世纪，且自公元 1883 年正式开馆以来从未因任何原因关闭过。

Architectural Terms

1. **square** /skweə(r)/ *n.* [C] an area of approximately square-shaped land in a city or a town, often including the buildings that surround it（城镇中的）广场（常包括周围的建筑）
   - 例句：The house is located in one of the city's prettiest *squares*.
     该房子位于市区最漂亮的广场之一。
   - 近义词：
     plaza /'plɑːzə/ *n.* [C] an open area or square in a town, especially in Spanish-speaking countries（尤指西班牙语国家城镇中的）露天广场

2. **mausoleum** /ˌmɔːsəˈliːəm/ *n.* [C] a building in which the bodies of dead people are buried 陵墓

- 例句：On their way back to the hotel, the journalists also visited that *mausoleum*.
  返回宾馆的途中，记者们还参观了那座陵墓。
- 近义词：
  tomb /tuːm/ *n.* [C] a large stone structure or underground room where someone, especially an important person, is buried（尤指重要人物的）坟墓；陵墓；冢

Polysemous Words

单词	常用词义	本文词义
since	*conj.* 自从……，从……以来	*conj.* 因为，既然

Difficult Sentences

1. Standing in Red Square, you can see the most significant buildings in the capital.

   Sentence analysis

   1) *standing in Red Square* 是现在分词形式作状语，其逻辑主语是主句的主语 *you*。

   2) *the most significant* 是 *significant* 的最高级形式，修饰 *buildings*。

   Translation

   站在红场上，你可以看到首都最重要的一些建筑。

2. The east side of the fortification was vulnerable since it was not protected by the rivers nor any other natural barriers.

   Sentence analysis

   since it was not protected by the rivers nor any other natural barriers 是 *since* 引导的原因状语从句。*since* 引导原因状语从句时表示已知的、显然的理由，但是 *since* 的语气比 *because* 弱，一般将其翻译为“既然”。

   Translation

   防御工事的东侧很脆弱，因为它没有受到河流或任何其他自然屏障的保护。

Text C

Egyptian Pyramids

Built during a time when Egypt was one of the richest and most powerful civilizations in the world, **the pyramids**—especially the Great Pyramids of Giza—are some of the most magnificent man-made structures in history. Their massive scale reflects the unique role that the pharaoh, or king, played in ancient Egyptian society.

No pyramids are more celebrated than the Great Pyramids of Giza, located on a **plateau** on the west bank of the Nile River, on **the outskirts** of modern-day Cairo. The oldest and largest of the three pyramids at Giza, known as the Great Pyramid, is the only surviving structure out of the famed Seven Wonders of the Ancient World. It was built for Pharaoh Khufu. The sides of the pyramid's base average 755.75 feet (230 meters), and its original height was 481.4 feet (147 meters), making it the largest pyramid in the world.

New Words

单词	音标	词义
civilization	/ˌsɪvəlaɪ'zeɪʃ(ə)n/	*n.* [C or U]（特定时期特定社会或国家的）文明，文化；文明社会
massive	/'mæsɪv/	*adj.* 巨大的；大量的
scale	/skeɪl/	*n.* [S or U] 大小，规模，范围
unique	/ju'niːk/	*adj.* 独一无二的，与众不同的；罕见的
celebrated	/'selɪbreɪtɪd/	*adj.*（尤指因品质或能力而）著名的，驰名的
famed	/feɪmd/	*adj.*（因被人们熟知而）著名的，闻名的
wonder	/'wʌndə(r)/	*n.* [C usually plural] 奇迹，奇观 *v.* [I or T] 疑惑，想知道

续表

单词	音标	词义
ancient	/'eɪnʃənt/	*adj.* 古代的；古老的；年代久远的
original	/ə'rɪdʒən(ə)l/	*adj.* 原来的，起初的；原件的，真迹的

Phrases

词组	词义
play a role in	在……起作用；扮演角色

Background Information

Nile River

尼罗河流域南起东非高原，北至地中海沿岸，东倚埃塞俄比亚高原，西邻刚果盆地，沿红海向西北延伸，流经非洲东部与北部，自南向北注入地中海。尼罗河长6670公里，是世界上最长的河流，与中非地区的刚果河以及西非地区的尼日尔河并列为非洲最大的三个河流系统。尼罗河流域是世界文明发祥地之一，这一地区的人民创造了灿烂的文化，突出的代表就是古埃及，古埃及语中尼罗河的意思是“大河”。

Cairo

开罗位于埃及的东北部，是埃及首都及该国最大的城市，也是非洲及阿拉伯世界最大的城市。开罗横跨尼罗河，是整个中东地区的政治、经济、文化和交通中心。

Architectural Terms

1. **the Pyramids** /'pɪrəmɪdz/ *n.* [plural] stone structures in Egypt of a pyramid shape that were built in ancient times as places to bury important people, especially pharaoh and queens（古埃及为埋葬重要人物特别是法老或王后修建的）金字塔

   - 例句：We set off to see *the Pyramids* and Sphinx.
     我们出发去看金字塔和狮身人面像。
   - 词源：pyramid 意为“棱锥体”，经希腊语、拉丁语在中世纪传入英语。汉语中的“金字塔”一词最早出现于普鲁士的传教士罗存德（Wilhelm Lobscheid）在 1866—1869 年陆续出版的四卷本《英华字典》（*English and Chinese Dictionary, with the Punti and Mandarin Pronunciation*），当时多指称中国传统建筑样式或地方名物。

2. **plateau** /'plætəʊ/ *n.* [C] a large flat area of land that is high above sea level 高原

   - 例句：The Qinghai-Xizang *Plateau* has long been known as the roof of the world.
     青藏高原素有“世界屋脊”之称。
   - 词源：plateau 来源于古法语中的“plat”一词，后又在古法语中演变为“platel”一词，经现代法语“plateau”进入英语。

3. **the outskirts** /'aʊtskɜːts/ *n.* [C] the areas that form the edge of a town or city 市郊，郊区（强调位置，通常指市区以外的地方）

   - 例句：The factory is in *the outskirts* of New Dlehi.
     这家工厂在新德里郊区。
   - 近义词：
     the suburbs /'sʌbɜːbs/ *n.* [C] the outer area of a town, rather than the shopping and business centre in the middle（与市区相接或相邻的）郊区；城外

Polysemous Words

单词	常用词义	本文词义
bank	*n.* [C] 银行	*n.* [C] 堤，岸

Difficult Sentences

1. Built during a time when Egypt was one of the richest and most powerful civilizations in the world, the pyramids—especially the Great Pyramids of Giza—are some of the most magnificent man-made structures in history.

   Sentence analysis

   1) *built during a time* 是过去分词作状语，其逻辑主语是主句的主语 *the pyramids*。

   2) *when Egypt was one of the richest and most powerful civilizations in the world* 是由关系副词 *when* 引导的定语从句，修饰先行词 *time*。

   Translation

   金字塔，特别是吉萨金字塔，是历史上最宏伟的人造建筑之一，建于埃及文明是世界上最富有、最鼎盛文明之一的时期。

2. Their massive scale reflects the unique role that the pharaoh, or king, played in ancient Egyptian society.

   Sentence analysis

   that the pharaoh, or king, played in ancient Egyptian society 是定语从句，由关系代词 *that* 引导。*that* 在定语从句中起引导定语从句、担任定语从句宾语、代替先行词 *role* 的作用。

   Translation

   它们（金字塔）庞大的规模反映了法老，即埃及国王，在古埃及社会中所扮演的独特角色。

Quiz for Chapter 2

Major: ______________ No.: ______________ Name: ______________

1 VOCABULARY AND GRAMMAR

A Check the best meaning for the underlined word in each sentence.

1) Known as the Outer Court, the southern ***portion*** of the Forbidden City features three main halls.

A. apportion B. part C. portfolio D. apart

2) Standing in Red Square, you can see the most ***significant*** buildings in the capital: the Kremlin, GUM department store, the State Historical Museum, Lenin's Mausoleum, and of course, St Basil's Cathedral.

A. notable B. inevitable C. interesting D. large

3) The pyramids were built during a time when Egypt was one of the most ***influential*** civilizations in the world.

A. original B. ancient C. beautiful D. powerful

4) These structures designated specific features to reflect the authority and ***status*** of the emperor.

A. stability B. state C. position D. venue

B Choose the right word.

5) Highways ______________ in all directions.

A. radiate B. shoot C. clear D. build

6) The dove is ______________ of peace.

A. realistic B. unique C. massive D. symbolic

7) Ladakh, ______________ more than 11,000 feet above sea level, is a high altitude desert where rainfall is usually very small.

A. situated B. have situated C. situates D. situate

8) The unique role ______________ Red Square played in the history of Russia is significant.

A. whose B. that C. who D. where

9) ____________ in this place, you can see the famous department store and the only museum of this city.

A. Stand B. Stood C. Standing D. Stands

2 COMMON KNOWLEDGE

10) The Imperial Palace is also called "____________".

A. Red Square B. the Outer Court
C. Beautiful Square D. the Forbidden City

11) The old Russian word for "beautiful" and "red" was the same, So Red Square means "____________".

A. Beauty Capital B. Beautiful Square
C. Soviet Russia D. the Great Pyramids of Giza

12) The Great Pyramids of Giza was built for ____________.

A. Pharaoh Khufu B. Giza
C. Egypt D. Greek

13) The Outer Court is the ____________ portion of the Forbidden City.

A. eastern B. western C. southern D. northern

14) Red Square is the largest and most famous square in ____________.

A. St. Petersburg B. Moscow
C. Egypt D. Cairo

15) It cost ____________ years to build the Forbidden City.

A. 6 B. 20 C. 4 D. 15

16) Cairo is the capital of ____________.

A. Hungary B. Russia C. Egypt D. Nile

Word Power

Major: ______________ No.: ______________ Name: ______________

Write the words and phrases under the pictures.

a plateau	an emperor	a square
the Pyramids	river banks	the outskirts

1.

2. ______________________

3. ______________________

4. ______________________

5. ______________________

6. ______________________

Group Work

Scan the webpage and complete the chart.

Try to find more information about famous squares in the world.

Famous Squares	Useful Information
Tiananmen Square	Tips: 1. It covers a total area of ________ 2. The meaning of the name is ________ 3. ________ 4. ________
Syntagma Square	Tips: 1. It is located in ________ 2. It is also called ________ 3. ________ 4. ________
________ **Square**	Tips: 1. ________ 2. ________ 3. ________ 4. ________
________ **Square**	Tips: 1. ________ 2. ________ 3. ________ 4. ________

Critical Thinking

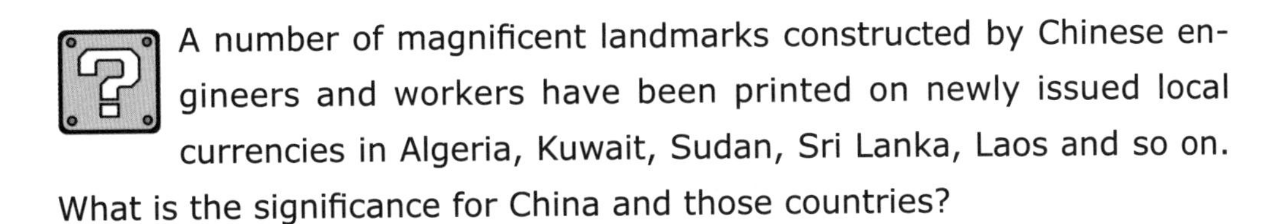

A number of magnificent landmarks constructed by Chinese engineers and workers have been printed on newly issued local currencies in Algeria, Kuwait, Sudan, Sri Lanka, Laos and so on. What is the significance for China and those countries?

For reference:

Plants with strong roots grow well, and efforts with the right focus will ensure success.

万物得其本者生，百事得其道者成。

——Liu Xiang

(A litterateur in the Han Dynasty)

A single flower does not make spring; one hundred flowers in full blossom bring spring to the garden.

一花独放不是春，百花齐放春满园。

—— The Wisdom of Ancient Aphorisms

Appreciation

Parthenon Temple
帕特农神庙
the landmark of Greece

Taj Mahal
泰姬陵
the landmark of India

Burj Al Arab
迪拜帆船酒店
the landmark of Dubai

Twin Towers
吉隆坡双子塔
the landmark of Kuala Lumpur

Self Assessment for Chapter 2

I can...

Very well OK A little

□ □ □ Use the architectural words in this part

□ □ □ Know the features of some architecture

□ □ □ List the landmarks of some countries

□ □ □ Recognize some famous architecture

Chapter 3

Modern Architecture

Preview:

By simply reviewing the path of Chinese architecture, this chapter will discuss the creative concept of Chinese modern architecture in the 21st century, and put forward to the idea of harmonious coexistence between man and nature.

Text A

National Stadium

Affectionately known as Bird's Nest, the National Stadium is situated in Olympic Park, Chaoyang District of Beijing City. It was designed as the main **stadium** of 2008 Beijing Olympic Games. Since October 2008, after the Olympics ended, it has been opened as a tourist attraction. Now, it's the center of international or domestic sports competition and recreation activities. In 2022, the opening and closing ceremonies of another important sport event, Winter Olympic Games was held here.

The form of the stadium looks like a big nest which embraces and nurses human beings. Also it looks rather like a cradle bearing human beings' hope of the future.

The construction of the National Stadium started on December 24, 2003. In July 2004, the project was stopped temporarily due to the amendment of the design. On December 27 of the same year, the construction was resumed and finished in March, 2008. The gross cost of the whole project is 2,267 million Chinese yuan (about 350 million U.S. dollars).

New Words

单词	音标	词义
affectionately	/ə'fekʃənətli/	*adv.* 热情地；体贴地；亲切地
domestic	/də'mestɪk/	*adj.* 家庭的；国内的；驯养的

续表

单词	音标	词义
recreation	/ˌriːkriˈeɪʃn/	*n.* [C] 消遣；娱乐
embrace	/ɪmˈbreɪs/	*vt.* 拥抱，包含 *vi.* 拥抱
cradle	/ˈkreɪdl/	*n.* [C] 摇篮；发祥地；支架 *vt.* 抚育
temporarily	/ˈtemprərəlɪ/	*adv.* 暂时
amendment	/əˈmendmənt/	*n.* [U or C] 修正；改进
resume	/rɪˈzjuːm/	*v.* [I or T] 重新开始 *n.* [C] 个人简历

Phrases

词组	词义
be situated in	坐落于……
tourist attraction	旅游胜地
due to	归因于……

Background Information

Olympic Park

北京奥林匹克公园是 2008 年北京奥运会和残奥会的举办地，总面积为 2864 英亩（1159 公顷）。公园设计包含十个场馆、奥运村和其他配套设施，后来被改造成一个综合性的多功能活动中心，供公众使用。

Architectural Terms

stadium /ˈsteɪdiəm/ *n.* [C] a large closed area of land with rows of seats around the sides and often with no roof, used for sports events and musical performances 体育场

- 例句：The college has announced its intention to enlarge its *stadium*.
  该学院已宣布了扩建其体育馆的打算。
- 词源：stadium 来自希腊语中的“stadion”一词，后来希腊语中性名词后缀“-on”换成了拉丁语中性名词后缀“-um”。“stadion”本指一段距离（180 米左右），是传说中大英雄赫拉克勒斯一口气跑下来然后需要“站立”休息的距离。古希腊人崇尚竞技，在这个距离的旁边设置看台就成了体育场。

Polysemous Words

单词	常用词义	本文词义
nurse	*n.* [C] 护士，保姆	*vt.* 看护，照料

Difficult Sentences

1. The form of the stadium looks like a big nest which embraces and nurses human beings.

   Sentence analysis

   look like 意为“看起来像”；*nurse* 在句中名词作动词，解释成“看护、照料”；*which* 引导的限制性定语从句作 *a big nest* 的定语，解释成“一个拥抱和照顾人类的大型鸟巢”。

   Translation

   体育场的形状看起来像一个拥抱和照顾人类的大型鸟巢。

2. Also it looks rather like a cradle bearing human beings' hope of the future.

   Sentence analysis

   like a cradle bearing... 现在分词作定语，等于 *which bears*。

   bear 在句中作动词，解释成“承载着”；*bearing* 现在分词做后置定语，解释成“一个承载着人类未来希望的摇篮”。

   Translation

   它看起来更像一个承载着人类未来希望的摇篮。

Text B

Hong Kong–Zhuhai–Macao Bridge

Hong Kong-Zhuhai-Macao Bridge (HZMB), the major construction work on the world's cross-sea bridge, which connects Zhuhai in Guangdong Province with Hong Kong and Macao, was completed on Oct 23, 2018. The bridge is built to meet the demand of passenger and freight land transport among the Chinese Mainland, the Hong Kong Special Administrative Region and the Macao Special Administrative Region, and to enhance the economic and sustainable development of the three places.

The bridge is a 55-kilometre bridge-tunnel system. The HZMB, which is located at the waters of the Lingdingyang of Pearl River Estuary, is a huge sea crossing, linking the Hong Kong SAR, Zhuhai city and the Macao SAR.

The project is made up of a 22.9-kilometer-long main bridge, a 6.7-kilometer-long **tunnel** and an artificial island off the bridge. It will bring people in Hong Kong, Macao and Guangdong within a "one-hour living circle", which is expected to attract more visitors to the Guangdong-Hong Kong-Macao Greater Bay Area.

To build the bridge, China Railway Shanhaiguan Bridge Group purchased two loaders, **costing** up to 130 million Chinese yuan (nearly 20 million U.S. dollars) each.

New Words

单词	音标	词义
freight	/freɪt/	*n.* [U] 货运；运费；货物

续表

单词	音标	词义
transport	/'trænspɔ:t/	*n.* [U] 运输；运输机
enhance	/ɪn'hæns/	*vt.* 提高；加强；增加
economic	/ˌi:kə'nɒmɪk/	*adj.* 经济的，经济上的；经济学的
sustainable	/sə'steɪnəbl/	*adj.* 可持续的
estuary	/'estʃuəri/	*n.* [C] 河口；江口；海湾
artificial	/ˌɑ:tɪ'fɪʃl/	*adj.* 人造的；虚伪的
bay	/beɪ/	*n.* [C] 海湾；月桂；分隔间

Phrases

词组	词义
be opened to the public	对公众开放
meet the demand of	满足需求
sustainable development	可持续发展
be made up of	由……组成

Background Information

Lantau Island

大屿山是中国香港最大的岛屿，面积为141.6平方公里，主峰凤凰山海拔935米，是全香港第二高峰。这里海岸线漫长曲折，港湾与沙滩、高山与流水、自然景观与历史古迹交相辉映。山上气势磅礴，有“凌绝顶”之称。山下有罗汉寺，寺内的罗汉洞及罗汉泉景色迷人。山的西面有宝莲寺和“天坛大佛”，北面有古堡，东南海岸则有香港海岸最长的海水浴场——长沙湾渔场。

Pearl River Delta

珠江三角洲，简称珠三角，是西江、北江共同冲积成的大三角洲与东江冲积成的小三角洲的总称。珠江三角洲位于广东省的珠江下游，毗邻港澳，与东南亚地区隔海相望，海陆交通便利，被称为中国的“南大门”。珠江三角洲是中国人口集聚最多、创新能力最强、综合实力最强的三大城市群之一。世界银行于 2015 年 1 月 26 日发布的报告显示，珠江三角洲已超越日本东京，成为世界人口最多和面积最大的城市群。

Architectural Terms

1. **tunnel** /'tʌnəl/

   ➢ *n.* [C] a long passage under or through the ground, especially one made by people 隧道；地道；坑道

   - 例句：The results provide references for the anti-seismic *tunnel* design.
     此结论为隧道抗震设计提供了依据。

   ➢ *v.* [I or T] to dig a tunnel 挖通道或隧道

   - 例句：The decision has not yet been made whether to *tunnel* under the river or build a bridge over it.
     还没有决定是在河底挖掘隧道还是在河上建桥。

2. **cost** /kɒst/

   ➢ *n.* [U] the amount of money that is needed in order to buy, do, or make it 费用

   - 例句：Painted walls look much more interesting and don't *cost* much.
     粉刷过的墙壁看上去更有趣，成本也不高。

   ➢ *vt.* to calculate the future cost of something 计算成本；计算花费

   - 例句：Has your plan been properly *costed* (out)?
     你的方案好好计算过成本吗?

   - 近义词：

✧ charge /tʃɑːdʒ/ *n.* [C or U] the amount of money that you have to pay for something, especially for an activity or service 收费；价格

● 例句：Is there a *charge* for children or do they go free?
小孩要收费还是免费？

✧ expense /ɪk'spens/ *n.* [C] the use of money, time, or effort 花钱；支付

● 例句：Buying a bigger car has proved to be well worth the *expense*.
事实证明，买一个更大的汽车是值得的。

Difficult Sentences

1. The HZMB, which is located at the waters of the Lingdingyang of Pearl River Estuary, is a huge sea crossing, linking the Hong Kong SAR, Zhuhai city and the Macao SAR.

Sentence analysis

which is located at the waters of the Lingdingyang of Pearl River Estuary 是非限制性定语从句。

Translation

超大型跨海交通工程港珠澳大桥跨越珠江口伶仃洋，东接香港特别行政区，西接广东省珠海市和澳门特别行政区。

2. It will bring people in Hong Kong, Macao and Guangdong within a "one-hour living circle", which is expected to attract more visitors to the Guangdong-Hong Kong-Macao Greater Bay Area.

Sentence analysis

which 引导的是非限制性定语从句。

Translation

港珠澳大桥的建成，令粤港澳地区的居民形成了"一小时生活圈"，有望吸引更多游客前往粤港澳大湾区。

Text C

New Suzhou Museum

Founded in 1960 and originally located in the national historic landmark, Suzhou Museum has been a highly-regarded regional museum with a number of significant Chinese cultural relics.

Under the design concept of "Chinese style with innovation, Suzhou style with creativity", the museum was built to be a modern, artistic and comprehensive museum in terms of its selected site and quality construction. Not only does it have the characteristics of a garden of Suzhou style, but also contains the simple **geometric** balance of the modern art as well as the exquisite structural layout in full function.

The museum is said to be the last design of I.M. Pei in his career. Therefore, not only does it become a monumental building in Suzhou, but also a significant construction, merging the traditional Chinese architectural design with the future. It enhances the protection of Suzhou cultural heritages, and enables Suzhou Museum to turn on a new page.

New Words

单词	音标	词义
highly-regarded	/'haɪli rɪ'gɑːdɪd/	*adj.* 高度关注的，很受关注的
relic	/'relɪk/	*n.* [C] 神圣的遗物；遗迹；纪念物
innovation	/ˌɪnə'veɪʃn/	*n.* [C or U] 新观念；革新
creativity	/ˌkriːeɪ'tɪvəti/	*n.*[U] 创造力，创造性
comprehensive	/ˌkɒmprɪ'hensɪv/	*adj.* 综合的；有理解力的
merge	/mɜːdʒ/	*v.*[I or T] (使) 合并；(使) 融合

Phrases

词组	词义
under the concept of	在……理念下
in terms of	根据，就……而言
be said to	据说

Background Information

New Suzhou Museum

苏州博物馆新馆的建筑造型与所处环境自然融合，空间处理独特，建筑材料考究，最大限度地把自然光线引入到室内。在建筑构造上，玻璃、钢铁结构让现代人可以在室内借到大片天光。屋面形态的设计突破了中国传统建筑“大屋顶”在采光方面的束缚。屋顶之上立体几何形体的玻璃天窗设计独特，借鉴了中国传统建筑中老虎天窗的做法并进行改良，天窗开在了屋顶的中间部位，这样的立体几何形天窗和其下的斜坡屋面形成一个折角，呈现出三维造型效果，不仅解决了传统建筑在采光方面的实用性难题，更丰富和发展了中国建筑的屋面造型样式。

Architectural Terms

geometric /ˌdʒɪːə'metrɪk/ *adj.* a geometric pattern or arrangement is made up of shapes such as squares, triangles, or rectangles 几何学的，几何学图形的

- 例句：The cube is a solid *geometric* figure.

  立方体是一种牢固的几何图形。

- 联想词汇：

✧ geometrician /ˌdʒɪəme'trɪʃən/ *n.* [C] a mathematician specializing in geometry 几何学家

Difficult Sentences

1. Not only does it have the characteristics of a garden of Suzhou style, but also contains the simple geometric balance of the modern art as well as the exquisite structural layout in full function.

   Sentence analysis

   not only 是句首倒装；*as well as* 连词解释为“此外，除此之外”。

   Translation

   它既有苏州园林的特色，又包含了现代艺术简约的几何平衡，以及功能齐全的精致结构布局。

2. Therefore, not only does it become a monumental building in Suzhou, but also a significant construction, merging the traditional Chinese architectural design with the future.

   Sentence analysis

   not only 是倒装结构；*merging* 作伴随状语，与谓语动词同时发生，一般使用分词形式表示。

   Translation

   因此，它不仅成为苏州的一座纪念性建筑，也是一座重要的建筑，将中国传统建筑设计与未来融为一体。

3. It enhances the protection of Suzhou cultural heritages, and enables Suzhou Museum to turn on a new page.

   Sentence analysis

   此句用两个动词 *enhance* 和 *enable* 连接句子，首尾对应；*cultural heritage* 解释为“文化遗产”。

   Translation

   它加强了对苏州文化遗产的保护，使苏州博物馆翻开新的一页。

Quiz for Chapter 3

Major: ______________ No.: ______________ Name: ______________

1 VOCABULARY AND GRAMMAR

A Check the best meaning for the underlined word in each sentence.

1) The form of the stadium looks like a big nest which embraces and ***nurses*** human beings.

A. take care of B. doctors C. friends D. teachers

2) Suzhou Museum has been a ***highly-regarded*** regional museum with a number of significant Chinese cultural relics.

A. recognized B. organized C. supported D. helped

3) Therefore, not only does it become a ***monumental*** building in Suzhou, but also a significant construction, merging the traditional Chinese architectural design with the future.

A. significant B. stone C. rock D. heaven

4) Now, it's the center of international or domestic sports competition and ***recreation*** activities.

A. amusement B. competition C. pleasure D. creation

B Choose the right word.

5) The project is __________ a 22.9-kilometer-long main bridge, a 6.7-kilometer-long tunnel and an artificial island off the bridge.

A. made up of B. make of C. look at D. keep from

6) Not only __________ have the characteristics of a garden of Suzhou style, but also contains the simple geometric balance of the modern art as well as the exquisite structural layout in full function.

A. dose it B. it does C. it was D. was it

7) Under the design concept of "Chinese style with innovation, Suzhou style with creativity", the museum __________ to be a modern, artistic and comprehensive museum.

A. was built B. was building C. building D. built

8) To build the bridge, China Railway Shanhaiguan Bridge Group purchased two loaders, __________ up to 130 million yuan.

A. cost B. costing C. to be cost D. to cost

2 COMMON KNOWLEDGE

9) In 2022, the opening and closing ceremonies of another important sport event, Winter Olympic Games was held in __________.

A. Beijing B. Nanjing C. Hangzhou D. Shanghai

10) As the world's longest seabased project, __________ has been named as one of the "seven wonders of the modern world" by The Guardian.

A. HZMB B. Yangtze River Bridge

C. Hangzhou Bay Bridge D. Jiashao Bridge

11) Where is the National Stadium?

A. In Pudong New District. B. In Haidian District.

C. In Chaoyang District. D. In Dongcheng District.

12) When was the National Stadium completed?

A. 2007. B. 2008. C. 2009. D. 2010.

13) When did the Hong Kong-Zhuhai-Macao Bridge open to the public?

A. 2018. B. 2017. C. 2016. D.2015.

14) The new Suzhou Museum was designed by the famous architect ______.

A. I.M. Pei B. Antoine Predock C. Ben van Berkel D. Li Xinggang

15) __________ is the English abbreviation for "Hong Kong-Zhuhai-Macao Bridge"?

A. HZMB B. HSMB C. ZHMB D. ZSMB

Word Power

Major: ________________ No.: ________________ Name: ________________

Write the words and phrases under the pictures.

estuary	stadium	relics
geometric	monument	tunnel

1.________________________

2.________________________

3.________________________

4.________________________

5.________________________

6.________________________

Group Work

Scan the webpage and complete the chart.

Try to find more information about modern architecture.

Modern Architecture	Useful Information
	Tips: 1. Name: ________ 2. Designer: ________ 3. Details : ________
	Tips: 1. Name: ________ 2. Designer: ________ 3. Details : ________
	Tips: 1. Name: ________ 2. Designer: ________ 3. Details : ________
	Tips: 1. Name: ________ 2. Designer: ________ 3. Details : ________

Critical Thinking

In order to adapt to the new situation, what are the characteristics of the modern architecture?

For reference:

We need to strive for both scientific and technological innovation and institutional innovation, build synergy between market and technology, and help bring to fruition new technologies, new business forms and models to fully unlock our development potential.

我们要推动科技创新和制度创新两个轮子一起转，市场和技术和谐共振，让新技术、新业态、新模式不断开花结果，最大限度释放发展潜能。

——the 25th anniversary APEC Economic Leaders' Meeting

Innovation is to create a resource.

创新就是创造一种资源。

—— Peter F. Drucker

(the Father of Modern Management)

Appreciation

Beijing Daxing International Airport
北京大兴国际机场

Tianjin Library
天津图书馆

The Oriental Pearl Radio & TV Tower
东方明珠广播电视塔

Raffles City Hangzhou
杭州来福士广场

Self Assessment for Chapter 3

I can...

Very well OK A little

☐ ☐ ☐ Use the architectural terms in this part

☐ ☐ ☐ Master some information of modern architecture

☐ ☐ ☐ Say something about modern architecture

☐ ☐ ☐ Recognize some modern architecture

Chapter 4

Famous Architects

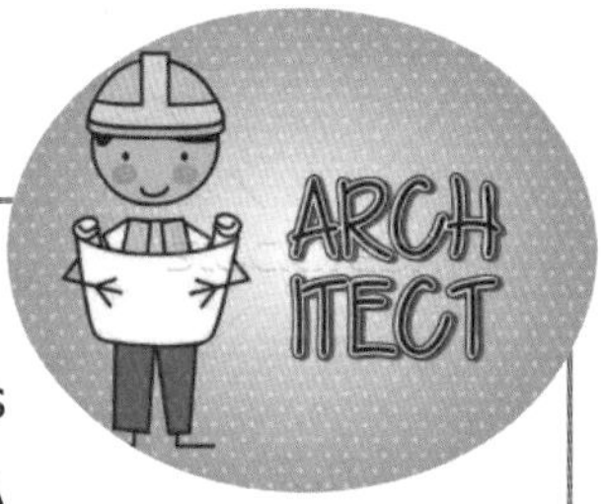

Preview:

An architect is a person who plans, designs and oversees the construction of buildings. A great number of architects make significant contributions to the built environment and humanity through their projects. In this chapter, we will go through three famous modern architects and appreciate their architectural style and their typical works.

Text A

Wang Shu

Wang Shu is a Chinese architect. He is the dean of the School of Architecture of the China Academy of Art. In 2012, Wang Shu became the first Chinese citizen to win the Pritzker Architecture Prize, he creates modern buildings making use of traditional materials and applying older techniques. Wang Shu is a keen supporter of architectural heritage where globalization has stripped cities of their special attributes.

In 2008, he completed the Ningbo Museum. The design is a conceptual combination of mountains, water and oceans, as the East China Sea has played an important role in the history of Ningbo. Features of Jiangnan residences are integrated into the museum design by decorations made from old tiles and bamboo. The building's facade is constructed entirely of recycled bricks salvaged from buildings which had been **demolished** to facilitate new developments, and its shape—resembling nearby mountains—reflects its natural **setting**. The museum won the 2009 Lu Ban Prize, the top architecture prize in China.

New Words

单词	音标	词义
dean	/diːn/	*n.* [C] 院长；系主任；主持牧师
academy	/ə'kædəmi/	*n.* [C] 学院；学会；专科院校
keen	/kiːn/	*adj.* 强烈的；敏锐的，渴望的
strip	/strɪp/	*vt.* 除去；剥夺；拆卸 *n.* [C] 条，带；商业街；连环画
attribute	/ə'trɪbjuːt/	*n.* [C] 属性；特质 *vt.* 归属；把……归于
conceptual	/kən'septʃuəl/	*adj.* 概念（上）的；观念（上）的
decoration	/ˌdekə'reɪʃn/	*n.* [C] 装饰品；奖章 *n.* [U] 装饰； 装饰风格
tile	/taɪl/	*n.* [C] 瓷砖，瓦片 *vt.* 铺以瓦片；铺以瓷砖
salvage	/'sælvɪdʒ/	*vt.* 抢救；打捞；挽回 *n.* [U] 打捞；救助；抢救出的财货
facilitate	/fə'sɪlɪteɪt/	*vt.* 促进；帮助；使容易

Background Information

the China Academy of Art

中国美术学院创建于 1928 年，时称国立艺术院，现有杭州南山、杭州象山、杭州良渚、上海张江四大校区。中国美术学院设有中国画与书法艺术学院、雕塑与公共艺术学院、建筑艺术学院、艺术人文学院等二级学院，王澍为建筑艺术学院现任院长。

Architectural Terms

1. **demolish** /dɪ'mɒlɪʃ/ *vt.* to pull or knock down a building 拆毁；拆除（建筑物）
   - 例句：The factory is due to be *demolished* next year for the development of the city.
     为了城市的发展，这个工厂定于明年拆除。
   - Common demolision equipment:
   - ➢ crane 起重机
   - ➢ excavator 挖掘机
   - ➢ bulldozer 推土机
   - ➢ rotational hydraulic shears 旋转液压剪

2. **setting** /'setɪŋ/ *n.* [C] a set of surroundings 环境
   - 例句：Working in a rural *setting*, he is inclined to design the works with idyllic style.
     在乡村环境中工作，他更喜欢设计田园风格的作品。
   - 近义词：
   - ✧ environment /ɪn'vaɪrənmənt/ *n.* [C] the natural world in which people, animals and plants live 自然环境；生态环境
   - ✧ surroundings /sə'raʊndɪŋz/ *n.* [plural] everything that is around 周围的环境

Difficult Sentences

1. He creates modern buildings making use of traditional materials and applying older techniques.

   Sentence analysis

   making use of... 是现在分词引导的宾语补足语，补充说明 *modern buildings*。

   Translation

   他创作了一系列利用传统材料且运用古老技术的现代建筑。

2. Wang Shu is a keen supporter of architectural heritage where globalization has stripped cities of their special attributes.

   Sentence analysis

1) *where globalization ...attributes* 是 *where* 引导的定语从句，修饰 *architectural heritage*，表示全球化剥夺了一些城市的特质。

2) *keen supporter* 意思是"狂热的支持者"；*strip* 意思是"剥夺、剥离"。

Translation

王澍是建筑遗产的热烈拥护者，而一些建筑遗产却被全球化剥夺了一些城市的特质。

3. The design is a conceptual combination of mountains, water and oceans, as the East China Sea has played an important role in the history of Ningbo.

Sentence analysis

1) *as the East China Sea ... Ningbo* 中 *as* 引导原因状语从句，表示"因为"。

2) *conceptual combination* 意思是"概念上的组合"。

Translation

设计是山、水和海洋的概念组合，因为东海在宁波历史上扮演了重要的角色。

4. The building's facade is constructed entirely of recycled bricks salvaged from buildings which had been demolished to facilitate new developments.

Sentence analysis

1) *which had been demolished ... developments* 是定语从句，修饰 *buildings*，说明这些建筑物为了促进新发展已被拆除。

2) *recycled bricks* 意思是"回收的砖块"；*salvaged* 过去分词作补语，补充说明这些砖块是从那些已拆除的建筑物中回收的。

Translation

博物馆的正面是用为促进新发展而拆除的建筑物中回收的砖块建造的。

Text B

Zaha Hadid

Zaha Hadid was a British-Iraqi architect. She was the first woman to receive the Pritzker Architecture Prize in 2004. Hadid was categorized as a major figure in Architectural Deconstructivism and her work was also described as an example of futuristic architecture characterized by curving facades, sharp angles, and severe materials such as concrete and steel, but she did not regard herself as a follower of any style or school.

The London Aquatics Centre designed by Zaha Hadid is an indoor **facility** with two 50-meter swimming pools and a 25-meter diving pool. As one of the main venues of the 2012 Summer Olympics, it was used for the swimming, diving and synchronized swimming events. The building covers three swimming pools, and seats 17,500 spectators at the two main pools. The roof, made of steel and aluminum and covered with wood on the inside, rests on just three **supports**. It was the first 2012 Olympic building begun but the last to be finished. Jacques Rogge, IOC President, described the centre as a "masterpiece".

New Words

单词	音标	词义
categorize	/'kætəgəraɪz/	*vt.* 分类，把……加以归类
deconstructivism	/di:kən'strʌktivizəm/	*n.* [U] 解构主义
futuristic	/ˌfju:tʃə'rɪstɪk/	*adj.* 未来派的；未来主义的
characterize	/'kærəktəraɪz/	*vt.* 描绘……的特性；具有……的特征

续表

单词	音标	词义
aquatics	/ə'kwætɪks/	*n.* [U] 水上运动或表演
spectator	/spek'teɪtə(r)/	*n.* [C]（尤指体育比赛）观看者，观众
aluminum	/ə'lu:mɪnəm/	*n.* [U] 铝

Phrases

词组	词义
synchronized swimming	花样游泳

Background Information

Architectural Deconstructivism

解构主义建筑是在 20 世纪 80 年代晚期开始发展的后现代建筑，它的特别之处为破碎的想法和非线性设计的过程，在结构的表面和非欧几何上花功夫，形成建筑学设计原则的变形与移位。建筑的视觉外观利用不可预料和纷乱描绘的刺激形成了无数的解构主义“样式”。

Royal Ontario Museum

Dancing House in Prague

Walt Disney Concert Hall

Futuristic Architecture

未来主义建筑是 20 世纪初期起源于意大利的建筑形式，特点是反历史主义，多运用长线条，象征速度、运动、紧迫性和抒情性。

Nautilus Eco-Resort

Metropol Parasol

Jacques Rogge

雅克·罗格 (1942—2021)，比利时人，曾是比利时整形外科医生，体育强项是帆船、橄榄球。雅克·罗格 于 2001 年到 2013 年任国际奥委会主席。

Architectural Terms

1. **facility** /fə'sɪləti/

   ➢ n. [C] a place, especially including buildings, where a particular activity happens（尤指包含多个建筑物，有特定用途的）场所

   - 词组：nuclear research *facility* 核研究中心
     military *facility* 军事机构
     new sports *facility* 新的运动场所

   ➢ n. [plural] buildings, services, equipment, etc. that are provided for a particular purpose 设施；设备

   - 例句：The hotel has special *facilities* for welcoming disabled people.
     这家旅馆有专供残疾人使用的设施。
   - 近义词：

   equipment /ɪ'kwɪpmənt/ *n.* [U] the things that are needed for a particular purpose or activity 设备；器材

2. **support** /sə'pɔːt/ *n.* [C] a thing that holds sth. and prevents it from falling 支撑物；支承；支柱；支座

   - 例句：The *supports* under the bridge were starting to bend.
     桥下的支柱开始弯曲。
   - Function of a structural support:
     A structural support is a part of a building or structure that provides the necessary stiffness and strength in order to resist the internal forces (vertical forces of gravity and lateral forces due to wind and earthquakes) and guide them safely to the ground.

Polysemous Words

单词	常用词义	本文词义
severe	*adj.* 严厉的；剧烈的；苛刻的	*adj.* 坚固的，坚硬的
seat	*n.* [C] 座位	*vt.* 可坐……人；能容纳……人
rest	*n.* [C or U]& *v.* [I or T] 休息	*vt.* 以……为基础；支撑

Difficult Sentences

1. Her work was also described as an example of futuristic architecture characterized by curving facades, sharp angles, and severe materials such as concrete and steel.

Sentence analysis

1) *characterized* 是过去分词作后置定语，修饰 *futuristic architecture*，用以说明未来主义建筑的特点。

2) *curving facades* 意思是"弧形立面"；*sharp angles* 意思是"尖角"；*severe materials* 意思是"坚硬的材料"。

Translation

她的作品被认为是未来主义建筑的典范，其特点是多用弧形立面、尖角以及混凝土和钢筋等坚硬的材料。

2. The roof, made of steel and aluminum and covered with wood on the inside, rests on just three supports.

Sentence analysis

made of...inside 是过去分词作后置定语，修饰 *roof*，用以说明屋顶的材质。

Translation

屋顶由钢和铝制成，内部覆盖木头，仅靠三个支架撑起来。

Text C

Robert Venturi

Robert Venturi, an American architect, was one of the major architectural figures of the twentieth century. Venturi was awarded the Pritzker Architecture Prize in 1991.

The Vanna Venturi House designed for his mother is one of the prominent works of Robert Venturi. It is best known for its facade—a monumental **gable** with an oversized chimney in its centre and an assortment of mismatched windows. Inside, five rooms are arranged around a combined hearth and **staircase**. The living room is at the centre, with the dining space and separate kitchen on one side, the master bedroom and utility room on the other, and an **attic** bedroom located above. This building is full of contradictions: it is both complex and simple, open and closed, big and little; some of its elements are good on one level and bad on another. The house was constructed with intentional formal architectural, historical and aesthetic contradictions. Its influence is so great that it is now credited as the first typical work of the postmodern architecture movement.

New Words

单词	音标	词义
figure	/'figə(r)/	*n.* [C] 人物；数字；图形；体形；画像 *vt.* 计算；认为；描述 *vi.* 扮演角色；出现；卷入

续表

单词	音标	词义
prominent	/'prɒmɪnənt/	*adj.* 杰出的；突出的，显著的
oversize	/'əʊvəsaɪzd/	*adj.* 过大的，硕大的；超大号的
assortment	/ə'sɔːtmənt/	*n.* [C] 各种各样；混合物
mismatch	/ˌmɪs'mætʃ/	*vt.* 使错配；搭配不当
hearth	/hɑːθ/	*n.* [C] 炉床；壁炉地面；家和家庭生活
utility	/juː'tɪləti/	*n.* [C] 公共事业 *n.* [U] 实用；效用 *adj.* 实用的；通用的；有多种用途的
contradiction	/ˌkɒntrə'dɪkʃn/	*n.* [C or U] 矛盾；否认；反驳
element	/'elɪmənt/	*n.* [C] 元素；要素；原理；自然环境
intentional	/ɪn'tenʃənl/	*adj.* 故意的；蓄意的；策划的
aesthetic	/iːs'θetɪk /	*adj.* 美的；美学的；审美的 *n.* [C] 审美；美学

Background Information

Postmodern Architecture

后现代主义建筑是现代以后各流派建筑的总称，包含了多种风格的建筑。最早提出后现代主义的是美国建筑家罗伯特·文丘里，他提出“少即乏味”(Less is a bore)的看法，主张用历史建筑因素和美国的通俗文化赋予现代建筑审美性和娱乐性。美国建筑师罗伯特·斯特恩提出后现代主义建筑有三个特征：采用装饰；具有象征性或隐喻性；与现有环境融合。

BMW Museum in Munich, Germany

CCTV Headquarters in Beijing, China

Architectural Terms

1. **gable** /'geɪbl/ *n.* [C] the upper part of the end wall of a building, between the two sloping sides of the roof, that is shaped like a triangle 三角墙；山墙

   - 例句：A roof with two slopes that form a triangle at each end is called a *gable* roof.

     有两个斜面在两端各成一个三角形的屋顶，称为三角墙屋顶。

   - 比较：

     承重墙：structural wall，bearing wall

     非承重墙：partition wall

2. **staircase** /'steəkeɪs/ *n.* [C] a set of stairs inside a building including the posts and rails that are fixed at the side（建筑物内的）楼梯

   - 例句：The house has a superb *staircase* made from oak and marble.

     这座房子里有用橡木和大理石做成的一流的楼梯。

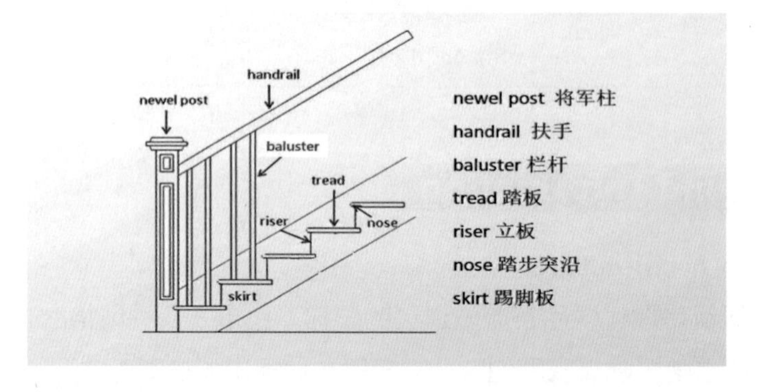

3. **attic** /'ætɪk/ *n.* [C] a room or space just below the roof of a house, often used for storing things（紧靠屋顶的）阁楼；顶楼

   - 例句：By converting the *attic*, they were able to have two extra bedrooms.

     通过改建阁楼，他们又多出了两间卧室。

   - 近义词：

   ✧ garret /'gærət/ *n.* [C] a room, often a small dark unpleasant one, at the top of a house, especially in the roof 阁楼；顶楼小屋

   ✧ loft /lɒft/ *n.* [C] a space just below the roof of a house, often used for storing things and sometimes made into a room 阁楼（常用以储物，或用作房间）

Polysemous Words

单词	常用词义	本文词义
separate	*vt.*（使）分开，分离；分割	*adj.* 独立的

Difficult Sentences

1. It is best known for its facade—a monumental gable with an oversized chimney in its centre and an assortment of mismatched windows.

   Sentence analysis

   it is best known for 意思是"因……而出名"；*an assortment of* 意思是"各种各样的"。

   Translation

   它最出名的是它的立面——一座巨大的三角墙，中间有一个超大的烟囱，还有各种各样不匹配的窗户。

2. The living room is at the centre, with the dining space and separate kitchen on one side, the master bedroom and utility room on the other, and an attic bedroom located above.

   Sentence analysis

   1) *with the dining space...above* 是补语，补充说明其他房间的位置。

   2) *the master bedroom* 意思是"主卧"；*utility room* 意思是"杂物间"。

   Translation

   客厅位于中央，一侧是用餐区和独立厨房，另一侧是主卧和杂物间，上方是阁楼卧室。

3. Its influence is so great that it is now credited as the first typical work of the postmodern architecture movement.

   Sentence analysis

   1) *so... that* 表示"如此……以至于"，*that* 引导结果状语从句。

   2) *is credited as* 意思是"被认为"。

   Translation

   它的影响如此之大，以至于被公认为是后现代主义建筑运动的第一个代表作。

Quiz for Chapter 4

Major: ______________ No.: ______________ Name: ______________

1 VOCABULARY AND GRAMMAR

A Check the best meaning for the underlined word in each sentence.

1) We're trying to get the medical ***facilities*** up and running again.

A. equipment B. factories C. furniture D. decoration

2) She was ***keen*** to learn about Chinese traditional culture when she came to Beijing.

A. emergent B. severe C. eager D. sensitive

3) ***Attic*** is a room or space just below the roof of a house, often used for storing things.

A. Venue B. Anteroom C. Support D. Loft

B Choose the right word.

4) We had to ____________ an old building in order to create a space with new functions.

A. reflect B. renovate C. resemble D. regard

5) It is a large commercial ____________, providing catering, entertainment and exhibition services.

A. gable B. anteroom C. venue D. complex

6) The App Store provided ____________ new apps free of charge at the beginning.

A. an assort of B. assortments of

C. an assortment of D. asserts of

7) He saw a piece of ____________ furniture near the table in the living room.

A. misstated B. mismatched C. misled D. mishit

8) We can't believe that his public speeches are in ____________ to his personal lifestyle.

A. assortment B. spectator C. masterpiece D. contradiction

9) The style of his works ____________ that of a famous painter in 19th century.

A. seems B. similar C. likes D. resembles

10) He's got himself into a dangerous situation ____________ he is likely to lose control over the plane.

A. what B. which C. where D. that

11) Any new source of energy will be very welcome, ____________ there is already a shortage of oil.

A. if B. as C. so D. although

12) The supermarket ____________ in the center of the city is a project of the most famous architect.

A. locating B. is located C. located D. is locating

2 COMMON KNOWLEDGE

13) Who is the first woman to win the Pritzker Architecture Prize?

A. Kazuyo Sejima. B. Jeanne Gang.
C. Zaha Hadid. D. Elizabeth Diller.

14) Which building was regarded as the first typical work of the postmodern architecture movement?

A. Louvre Pyramid. B. Vanna Venturi House.
C. London Aquatics Centre. D. Ningbo Museum.

15) What are the main materials chosen by Wang Shu in construction of Ningbo Museum?

A. Old tiles and recycled bricks. B. Glass and steel.
C. Steel and aluminum. D. Wood and concrete.

16) Which part was designed by the Chinese-American architect I.M. Pei in the Louvre Museum?

A. Square courtyard. B. Hall Napoléon.
C. The Pyramid entrance. D. Sully Wing.

Word Power

Major: ________ No.: ________ Name: ________

Write the words and phrases under the pictures.

categorize	hearth	decoration
demolish	tile and brick	utility room

1.________

2.________

3.________

4.________

5.________

6.________

Group Work

Scan the webpage and complete the chart.

Try to find more information about architectural design software.

Architectural Design Software	Main Functions and Features
SketchUp SketchUp	SketchUp is a 3D modeling computer program for ________________. 1. Visualize in no time 2. ________________ 3. ________________
AutoCAD AutoCAD	It is ________________(CAD) software that architects rely on to create ____________ drawings. 1. Industry-specific toolsets 2. ________________ experience across platforms 3. ________________ 4. ________________
Revit	Revit is a building information modeling software for architects. 1. ________________ 2. ________________ 3. ________________

Critical Thinking

In addition to the vocational skills and abilities, what else can you learn from those architects?

For reference:

We will build an educated, skilled, and innovative workforce, foster respect for model workers, promote quality workmanship, and see that taking pride in labor becomes a social norm and seeking excellence is valued as a good work ethic.

建设知识型、技能型、创新型劳动者大军，弘扬劳模精神和工匠精神，营造劳动光荣的社会风尚和精益求精的敬业风气。

—— The report of the 19th National Congress of CPC

I believe that architecture is a pragmatic art. To become art it must be built on a foundation of necessity.

我相信建筑是一种实用艺术。要成为艺术，它必须建立在必要性的基础上。

Design is something you have to put your hand to.

设计是需要通过实践来完成的。

——I.M. Pei

(A master of the modern architecture)

Appreciation

Fragrant Hill Hotel in Beijing
北京香山饭店
By I.M. Pei

Provincial Capitol Building in Toulouse, France
法国图卢兹省政府大楼
By Robert Venturi

Guangzhou Opera House
广州大剧院
By Zaha Hadid

Xiangshan Campus, China Academy of Art
中国美术学院象山校区
By Wang Shu

Self Assessment for Chapter 4

I can…

Very well OK A little

☐ ☐ ☐ Use the architectural words in this part

☐ ☐ ☐ Master some information of the architects

☐ ☐ ☐ Say something about modern architects

☐ ☐ ☐ Recognize some typical works of the architects

Chapter 5

Construction Companies

Preview:

While most people can think of a few amazing works of architecture that stand apart from the rest, many don't really think about the companies that helped to build them.

This chapter introduces three of the most impressive construction companies on the planet, and we'll learn a little bit about what sets each one apart from its peers.

Text A

China State Construction Engineering Corporation

China State Construction Engineering Corporation (CSCEC), founded in 1982, is one of the top 10 global construction companies. In 2020, CSCEC generated $234,425 million in revenue and moved up to 13th in the *Fortune* Global 500 list in 2021.

With seven listed companies and more than 100 secondary holding subsidiaries, CSCEC has been doing business in more than 100 countries and regions in the world, covering investment and development (real estate, construction financing and operation), construction engineering (housing and **infrastructure**) as well as **survey** and design (green construction, energy conservation and environmental protection, and e-commerce).

In China, CSCEC has built more than 90% of skyscrapers above 300 meters, three-quarters of key airports and satellite launch bases, one-third of urban utility tunnels and half of nuclear power plants. One out of every 25 Chinese lives in the house built by CSCEC.

Malacca Impression Opera Theatre

New Words

单词	音标	词义
corporation	/ˌkɔːpəˈreɪʃn/	*n.* [C] 公司；法人（团体）；社团
revenue	/ˈrevənjuː/	*n.* [U] 收入；财政收入；税收收入
fortune	/ˈfɔːtʃuːn/	*n.* [C] 财富；[U] 命运；运气
subsidiary	/səbˈsɪdiəri/	*n.* [C] 子公司 *adj.* 辅助的；附属的；子公司的

续表

单词	音标	词义
investment	/ɪn'vestmənt/	*n.* [C or U] 投资；投入
finance	/'faɪnæns/	*n.* [U] 财政，财政学；金融 *vi.* 筹措资金
conservation	/ˌkɒnsə'veɪʃn/	*n.* [U] 保存，保持；保护；节约
e-commerce	/i:kɒmərs/	*n.* [U] 电子商务

Phrases

词组	词义
listed company	上市公司
real estate	房地产
construction financing and operation	建设融资与运营
housing and infrastructure	住房和基础设施

Background Information

Fortune Magazine

《财富》是美国跨国商业杂志，总部设在纽约，由 Henry Luce 创办。该杂志定期发布排名名单，包括《财富》世界 500 强。《财富》世界 500 强也被称为全球 500 强，是按收入衡量的全球 500 强企业的年度排名。

Architectural Terms

1. **infrastructure** /ləʊ'keɪʃn/ *n.* [C or U] the basic systems and services that are necessary 基础设施；基础建设
   - 例句：Natural disasters can wreck a community's *infrastructure.*
     自然灾害可以破坏一个社区的基础设施。
   - 词源："infra-" 意为"低于，在下方"；structure 意为"结构"。

2. **survey** /'sɜːveɪ/ *n.* [C] the act of examining and recording the measurements, features, etc. of an area of land 测量；勘测；测绘

- 例句：The consultation of *survey* is the soul of engineering construction.
  勘察咨询是工程建设的灵魂。
- 近义词：
  investigation /ɪnˌvestɪ'geɪʃn/ *n.* [C or U] an official examination of the facts about a situation, crime, etc.（正式的）调查；侦查

Polysemous Words

单词	常用词义	本文词义
cover	*vt.* 掩盖；遮盖；覆盖	*vt.* 包括；包含；涉及；处理
house	*n.* [C] 住宅；某种用途的建筑物	*vt.* 给（某人）提供住处

Difficult Sentences

1. China State Construction Engineering Corporation (CSCEC), founded in 1982, is one of the top 10 global construction companies.

   Sentence analysis

   founded in 1982 是过去分词短语作状语，表示被动和完成。

   Translation

   中国建筑集团有限公司成立于 1982 年，是全球十大建筑企业之一。

2. One out of every 25 Chinese lives in the house built by CSCEC.

   Sentence analysis

   built by CSCEC 是过去分词短语作定语，修饰 *house*，表示被动，该短语可以改为定语从句 which is built by CSCEC。

   Translation

   每 25 个中国人中，就有 1 个住在中国建筑集团有限公司建造的房子里。

Text B

Power Construction Corporation of China

Power Construction Corporation of China (POWERCHINA) is a state-owned company **established** in 2009 and was ranked the 107th among the *Fortune* 500 enterprises in 2021.

POWERCHINA maintains a strong focus on contracting survey and design of hydraulic and water conservation projects. POWERCHINA is responsible for over 80 percent of the surveying, planning and designing of China's major hydropower projects. Involvement with projects within China includes the Three Gorges Project, and the Beijing-Shanghai high-speed railway.

Operating in over 102 countries, POWERCHINA has built more than 200 large and medium sized hydropower stations in various countries. POWERCHINA successfully completed some landmark projects, such as the Adama Wind Farm which became the largest wind farm in Sub-Saharan Africa and provided people with a promising future of renewable energy.

As one of the world's largest power construction enterprise committed to the construction of global energy source and infrastructures, POWERCHINA seeks to become the most competitive comprehensive construction group all over the world.

Zouxian Power Station

New Words

单词	音标	词义
enterprise	/'entəpraɪz/	*n.* [C] 公司；规划，事业

续表

单词	音标	词义
maintain	/meɪn'teɪn/	*vt.* 保持；维修；保养
contract	/'kɒntrækt; kən'trækt/	*n.* [C] 合同；婚约 *v.* [I or T] 收缩；订契约；订（婚）
hydraulic	/haɪ'drɒlɪk/	*adj.* 液压的；水力的；水力学的
involvement	/ɪn'vɒlvmənt/	*n.* [U] 参与；牵连；包含；混乱
renewable	/rɪ'nju:əbl/	*adj.* 可再生的；（合同、协议）可延长有效期的
competitive	/kəm'petətɪv/	*adj.* 竞争的；有竞争力的

Phrases

词组	词义
major hydropower projects	重大水电项目
hydropower station	水电站
be committed to	致力于……

Background Information

Three Gorges Project

三峡工程全称为长江三峡水利枢纽工程，整个工程包括一座混凝重力式大坝、泄水闸、一座堤后式水电站、一座永久性通航船闸和一架升船机。三峡工程建筑由大坝、水电站厂房和通航建筑物三大部分组成。大坝坝顶总长 3035 米，坝高 185 米，是迄今为止世界上规模最大的水利枢纽工程和综合效益最广泛的水电工程。

Adama Wind Farm

阿达玛风电场是埃塞俄比亚的第一个风电场，也是撒哈拉以南地区最大和非洲第二大的风电场。该风电场由中国和埃塞俄比亚企业共同建设，是中非可再生能源合作的标志性项目。

Architectural Terms

establish /ɪ'stæblɪʃ/ *vt.* to start or create an organization, a system, etc. that is meant to last for a long time 建立；创立；设立

- 例句：The building committee *established* last year shall meet monthly with the architect, contractors, etc.
  去年成立的这个建筑委员会将与建筑师、承包商等每月举行会议。
- 近义词：
  found /faʊnd/ *vt.* to bring something into existence 创立，建立；创办

Polysemous Words

单词	常用词义	本文词义
power	*n.* [U] 力量，能力	*n.* [U] 电力

Difficult Sentences

As one of the world's largest power construction enterprise committed to the construction of global energy source and infrastructures, POWERCHINA seeks to become the most competitive comprehensive construction group all over the world.

Sentence analysis

committed to the construction of global energy source and infrastructures 是过去分词短语作定语，修饰 *enterprise*，该短语可以改为定语从句 which is committed to the construction of global energy source and infrastructures。

Translation

作为致力于全球能源和基础设施建设的世界最大的电力建设企业之一，中国电力建设集团有限公司力求成为全球最具竞争力的综合性建设集团。

Text C

Gensler

Gensler

Founded in San Francisco in 1965, Gensler is a leading global design, planning, and consulting architecture firm with more than 5,500 professionals across the world working together to create more inclusive and resilient spaces that support the health and well-being of everyone.

As the firm's global footprint has grown, Gensler has launched megaprojects such as CityCenter Las Vegas, SFO Airport, and Shanghai Tower. Gensler delivers buildings across the global markets with a consistently high standard of service. As the world continues to emerge from the pandemic and economies begin to recover, many companies will find themselves poised for dramatic growth. Reimagining the future of cities as unprecedented challenges towards a critical yet hopeful time of transformation, Gensler is continually evolving the organization and **design processes** to deliver innovation aimed at helping clients and communities become more resourceful, resilient, and regenerative.

Shanghai Tower

As Gensler's founder Art Gensler put it, "we are a constellation of stars, with each individual playing a critical role in the company's overall success." With the mission to transform the built environment for a net zero future with every project, in every location, and for every person, Gensler is committed to creating a better world through the power of design.

New Words

单词	音标	词义
consulting	/kən'sʌltɪŋ/	*adj.* 顾问的，提供咨询的
inclusive	/ɪn'kluːsɪv/	*adj.* 包容广阔的；包含全部费用的
resilient	/rɪ'zɪliənt/	*adj.* 有弹力的；可迅速恢复的
well-being	/'wel biːɪŋ/	*n.* [U] 幸福安康
emerge	/ɪ'mɜːdʒ/	*vi.* 出现；摆脱；暴露
pandemic	/pæn'demɪk/	*n.* [C]（疾病）大规模流行的，广泛蔓延的
poise	/pɔɪz/	*vt.* 使平衡；准备（做某事） *vi.* 平衡，悬浮；准备好
dramatic	/drə'mætɪk/	*adj.* 巨大的；令人激动的；戏剧的
unprecedented	/ʌn'presɪdentɪd/	*adj.* 空前的；史无前例的
transformation	/ˌtrænsfə'meɪʃn/	*n.* [C or U]（彻底的）变化，转变
evolve	/ɪ'vɒlv/	*vt.* 发展；进化；使逐步形成；推断出 *vi.* 发展，进展；进化；逐步形成
community	/kə'mjuːnəti/	*n.* [C] 社区；[生态] 群落；团体
resourceful	/rɪ'sɔːsfl/	*adj.* 足智多谋的；资源丰富的
regenerative	/rɪ'dʒenərətɪv/	*adj.* 再生的
constellation	/ˌkɒnstə'leɪʃn/	*n.* [C] 星座；一系列（相关的想法、事物）；一群（相关的人）

Phrases

词组	词义
net zero	净零

Background Information

CityCenter Las Vegas

城中城，又称拉斯维加斯城中城，是一个位于美国内华达州赌城大道上的都会多用途设施，总楼地板面积达 1560500 平方米。从阿丽雅酒店眺望城中城七栋建筑中的四栋，由左至右分别为哈蒙酒店、水晶宫购物中心、维尔塔、文华东方酒店。

SFO Airport

旧金山国际机场（San Francisco International Airport）位于旧金山市以南大约 21 公里，是美国加州的一座大型商用机场，是旧金山湾区和北加州最大的机场和主要的国际门户。

Architectural Terms

design processes /dɪ'zaɪn'prəʊsesɪz/ a series of steps that architects use in creating architectural products 设计过程

- 例句：For architects, computer-aided design has become essential and has cheapened *design processes.*
  对于建筑师来说，计算机辅助设计已经变得至关重要，它能降低建筑设计过程成本。

Polysemous Words

单词	常用词义	本文词义
firm	*adj.* 坚定的；结实的；严格的	*n.* [C] 公司

Difficult Sentences

1. Founded in San Francisco in 1965, Gensler is a leading global design, planning, and consulting architecture firm with more than 5,500 professionals across the world working together to create more inclusive and resilient spaces that support the health and well-being of everyone.

   Sentence analysis

   1）在 *Gensler is a leading global design, planning, and consulting architecture firm* 中，*leading* 是现在分词，表示“领先的”，修饰后面的 *global design, planning, and consulting architecture firm*；而 *design, planning, and consulting*

是用逗号和连词 *and* 连接的三个并列项，文中用这个并列项修饰 *architecture firm*。

2）*with more than 5,500 professionals across the world working together to create more inclusive and resilient spaces that support the health and well-being of everyone* 是 *with*+ 名词短语 + 现在分词短语结构，该部分在本句子中作后置定语，修饰 *firm*。

3）*that support the health and well-being of everyone* 是 *that* 引导的定语从句，修饰前面的名词 *spaces*。

Translation

1965 年，Gensler 在旧金山成立。这是一家全球领先的设计、规划和咨询建筑公司。为了人类的健康和幸福，Gensler 的 5500 多名专业人员在世界各地共同努力创造更具包容性和更具活力的空间。

2. With the mission to transform the built environment for a net zero future with every project, in every location, and for every person, Gensler is committed to creating a better world through the power of design.

Sentence analysis

With the mission to transform the built environment for a net zero future with every project, in every location, and for every person 是 *with*+ 名词 + 不定式结构，该结构在本句子中作状语。 *is committed to* 的意思是“致力于……”，其中的 *to* 是介词，所以 *to* 后面的动词要用动名词形式。

Translation

Gensler 的使命是通过每个项目、每个地点和每个人来改变建筑环境，实现净零未来，从而通过设计的力量创造一个更美好的世界。

Quiz for Chapter 5

Major: ______________ No.: ______________ Name: ______________

1 VOCABULARY AND GRAMMAR

A Check the best meaning for the underlined word in each sentence.

1) In China, CSCEC has built more than 90% of ***skyscrapers***.
   A. houses B. modern buildings
   C. large houses D. very tall buildings

2) POWERCHINA was ranked the 107th among the *Fortune* 500 ***enterprises*** in 2021.
   A. schools B. companies C. organizers D. buildings

3) CSCEC is one of the top 10 global construction ***companies***.
   A. corporations B. components C. compositions D. comparisons

4) ***Founded*** in San Francisco in 1965, Gensler is a leading global design architecture firm.
   A. Found B. Developed C. Established D. Made

B Choose the right word.

5) Our new offices are still under ______________.
   A. build B. construction C. structure D. developing

6) POWERCHINA is ______________ for over 80 percent of China's major hydropower projects.
   A. waiting B. anxious C. eager D. responsible

7) CSCEC has been doing business in more than 100 countries, ______________ investment.
   A. to cover B. covers C. covering D. covered

8) One out of every 25 Chinese ______________ in the house built by CSCEC.
   A. live B. are living C. have lived D. lives

9) We are a constellation of stars, with each individual ______________ a critical role in the company's overall success.

A. played B. playing C. play D. plays

10) Gensler is continually ____________ the design processes to deliver innovation.

A. develop B. developed C. evolve D. evolving

11) POWERCHINA seeks to become the most ____________ construction group all over the world.

A. competed B. competition C. compete D. competitive

2 COMMON KNOWLEDGE

12) The *Fortune* Global 500 list is an annual ranking of the top 500 corporations worldwide as measured by ____________.

A. the prizes/awards B. the size of the company

C. the employees D. the revenue

13) POWERCHINA is a ____________ company.

A. private B. family C. state-owned D. school-owned

14) POWERCHINA maintains a strong focus on contracting survey and design of ____________ projects.

A. hydropower B. railway C. architecture D. skyscraper

15) Founded in San Francisco in 1965, Gensler is a leading global design, planning, and consulting ____________ firm.

A. agriculture B. accounting C. architecture D. aviation

16) Shanghai Tower is a tall green building designed by ____________.

A. BECHTEL B. Gensler C. POWERCHINA D. CSCEC

17) Gensler's founder is ____________.

A. the U.S. B. Bill Gates C. Art Gensler D. Trump

18) ____________ moved up to 13th in the *Fortune* Global 500 list in 2021.

A. BECHTEL B. Gensler C. POWERCHINA D. CSCEC

Word Power

Major: ________ No.: ________ Name: ________

Write the words and phrases under the pictures.

contract	infrastructure	skyscraper
real estate	revenue	wind farm

1.________

2.________

3.________

4.________

5.________

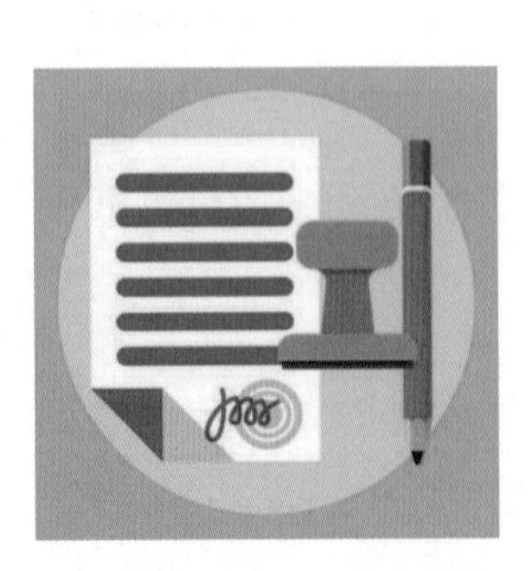

6.________

Group Work

Scan the webpage and complete the chart.

Try to find more information about famous construction companies.

Construction Companies	Useful Information
China Railway Group Limited	Tips: 1. Native name: ______ 2. Headquarters: ______ 3. Services: ______ 4. ______
China Communications Construction Company Limited	Tips: 1. Native name: ______ 2. Headquarters: ______ 3. Services: ______ 4. ______
______ **Construction company/Architects' firm**	Tips: 1. ______ 2. ______ 3. ______ 4. ______
______ **Construction company/Architects' firm**	Tips: 1. ______ 2. ______ 3. ______ 4. ______

Critical Thinking

What is the essential parts of the success of the entrepreneurs in the global top construction companies?

For reference:

In order to lead enterprises out of the immediate plight and toward a brighter future, entrepreneurs should promote entrepreneurship and seek self-improvement in terms of patriotism, innovation, integrity, social responsibility and global vision. Entrepreneurs should play a key role in creating a new development pattern in the new era, building the modern economic system and pushing for high-quality development.

企业家要带领企业战胜当前的困难，走向更辉煌的未来，就要在爱国、创新、诚信、社会责任和国际视野等方面不断提升自己，努力成为新时代构建新发展格局、建设现代化经济体系、推动高质量发展的生力军。

——Symposium with Entrepreneurs

When doing things, we should be honest not only to our consumers, but also to our company and employees. We should take the interests of others into consideration in everything we do.

做人做事讲诚信，不仅对客户、对消费者讲诚信，对公司和员工也要讲诚信，每做一件事都要考虑到别人的利益。

—— Dong Mingzhu

(The Chairman of Gree Electric Appliances, Inc. of Zhuhai)

Appreciation

Zhejiang Construction Engineering Group Co.,LTD.
浙江省建工集团有限责任公司

Zhejiang Province Institute of Architectural Design and Research
浙江省建筑设计研究院

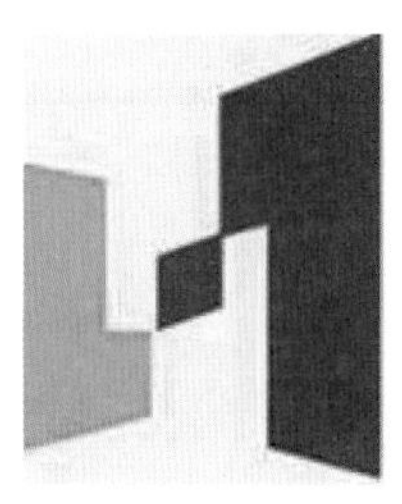

Zhejiang Academy of Building Research&Design Co.,LTD.
浙江省建筑科学设计研究院有限公司

Zhejiang Urban and Rural Planning Design Institute Co.,LTD.
浙江省城乡规划设计研究院有限公司

Self Assessment for Chapter 5

I can…

Very well	OK	A little	
□	□	□	Use the architectural words in this part
□	□	□	Master some information of construction companies
□	□	□	Say something about construction companies
□	□	□	Recognize some top companies' major projects

Chapter 6

Architecture Prizes

Preview:

Architecture prize plays a vital role for architectural culture. It is not only the direction of the architectural industry, but also an important reference for people outside the industry to make decisions on architectural projects and choose architects. In this chapter, we will go over three famous architecture prizes around the world to expand our horizons.

Text A

Pritzker Architecture Prize

The Pritzker Architecture Prize is an international award established by the Pritzker family through their Hyatt Foundation in 1979. It is often referred to as "the Nobel Prize of Architecture" and "the profession's highest honor."

The purpose of the Pritzker Architecture Prize is to honor a living **architect** or architects whose built work demonstrates a combination of talent, vision, and commitment; and has produced consistent and significant contributions to humanity and the built environment through the art of architecture.

The official ceremony granting the award takes place every year, usually in May, at an architecturally significant site throughout the world. The choice of **location** of the ceremony reinforces the importance of the built environment while providing a unique setting for the ceremony.

Many of the procedures and rewards of the Pritzker Prize are modeled after the Nobel Prize. Laureates of the Pritzker Architecture Prize receive a $100,000 grant, a formal citation certificate, and since 1987, a bronze medallion.

New Words

单词	音标	词义
demonstrate	/'demənstreɪt /	*vt.* 证明；展示；论证 *vi.* 示威
vision	/'vɪʒn/	*n.* [U] 眼光；远见；眼力
commitment	/kə'mɪtmənt/	*n.* [C or U] 忠诚；奉献；承诺，保证

consistent	/kən'sɪstənt/	*adj.* 一贯的；坚持的；始终如一的
humanity	/hjuː'mænəti/	*n.* [U] 人类；人性；仁慈
grant	/grɑːnt/	*vt.* （通常指官方）同意，准予，授予 *n.* [C] 拨款，补助金
ceremony	/'serəməni/	*n.* [C or U] 典礼；仪式
laureate	/'lɒriət/	*n.* [C] 获奖者，荣誉获得者
citation	/saɪ'teɪʃn/	*n.* [C] 嘉奖令；奖状；（法院的）传票
bronze	/brɒnz/	*adj.* 古铜色的，青铜色的
medallion	/mə'dæliən/	*n.* [C]（装饰用的）大奖牌，大纪念章

Phrases

词组	词义
a combination of	结合了……

Background Information

the Pritzker Family

Cindy & Jay Pritzker, Margot & Tom Pritzker

普利兹克家族的商业网络遍布全球，该家族通过其名下的凯悦基金会设立和赞助了普利兹克建筑奖，他们还因长期积极赞助各项教育、科学、医疗及文化活动而著称。

Hyatt

1957 年，创始人 Jay Pritzker 成立凯悦品牌。2004 年 12 月 31 日，普利兹克家庭企业利益集团拥有的全部酒店资产被整合至现在的凯悦酒店集团。

Architectural Terms

1．**architect** /'ɑːkɪtekt/ *n.* [C] a person whose job is to design new buildings and make certain that they are built correctly 建筑师

- 例句：The *architect* drew the house in section.
  建筑师绘制了房子的剖面图。
- 词源：architect 来源于希腊语中由名词“arkhos”（首领，统治者）和“tekton”（木匠，建造者）组成的复合词“arkhitekton”（主建造者），经拉丁语“architectus”、意大利语“architetto”、法语“architecte”发展演变成英语“architect”。

2．**location** /ləʊ'keɪʃn/ *n.* [C] a place or position 地点；位置

- 例句：The hotel is in a beautiful *location* overlooking the lake.
  该酒店的位置极佳，可以俯瞰整个湖面。
- 近义词：

✧ position /pə'zɪʃn/ *n.* [C] the place where something or someone is, often in relation to other things 位置；方位；地点

✧ site /saɪt/ *n.* [C] a place where something is, was, or will be built, or where something happened, is happening, or will happen（建筑物的）地点，位置；建筑工地；（某事发生的）地点，现场

Polysemous Words

单词	常用词义	本文词义
model	*n.* [C] 模范，榜样；模特；模型	*vt.* 模拟，模仿

Difficult Sentences

1．The Pritzker Architecture Prize is an international award established by the Pritzker family through their Hyatt Foundation in 1979.

Sentence analysis

established by the Pritzker family 是过去分词作后置定语，在句中表示被动。

一般情况下过去分词本身作定语的时候需要前置，只有当过去分词与副词、介词构成短语作定语的时候，需要后置。

Translation

普利兹克建筑奖是普利兹克家族名下的凯悦基金会于 1979 年创立的一个国际性奖项。

2. The purpose of the Pritzker Architecture Prize is to honor a living architect or architects whose built work demonstrates a combination of talent, vision, and commitment; and has produced consistent and significant contributions to humanity and the built environment through the art of architecture.

Sentence analysis

定语从句 *whose built work demonstrates a combination of talent, vision, and commitment* 修饰 *a living architect or architects*。

Translation

普利兹克建筑奖旨在表彰一位或多位当代建筑师在作品中所表现出的才智、眼光和责任感，以及他们通过建筑艺术对人文学科和建筑环境所做出的持续且显著的贡献。

3. The choice of location of the ceremony reinforces the importance of the built environment while providing a unique setting for the ceremony.

Sentence analysis

while providing a unique setting for the ceremony 作时间状语。

Translation

颁奖地点的选择凸显了建筑环境的重要性，同时也为典礼提供了独特的氛围。

Text B

AAA & ARCASIA

The ARCASIA Awards for Architecture (AAA) is the highest award of the Architects Regional Council Asia (ARCASIA), and also the highest architectural design award in Asia. ARCASIA was established in Jakarta, Indonesia, in 1979.

The organization holds the following objectives: to unite National Institutes of Architects on a democratic basis throughout the Asian **region**; to foster friendly, intellectual, artistic, educational and scientific ties; to foster and maintain professional contacts, mutual co-operation and assistance among member institutes; to represent architects of the member institutes at national and international levels; to promote the recognition of the architect's role in society; to promote the development and education of architects and the architectural profession in their service to society; and, to promote research and technical advancement in the field of the built environment.

The AAA aims at acknowledging exemplary work done by architects working in Asia. It encourages the sustenance of the Asia spirit, the development and improvement of the Asian construction environment and enhancement of the awareness of the role of architecture and architects in the socio-economic and cultural life of Asian countries.

Dongziguan affordable housing by Meng Haofan (2019 gold medal)

New Words

单词	音标	词义
council	/'kaʊnsl/	*n.* [C] 委员会；政务委员会；地方议会
unite	/ju'naɪt/	*v.* [I or T]（使）结合；（使）联合
democratic	/ˌdemə'krætɪk/	*adj.* 民主的
foster	/'fɒstə(r)/	*vt.* 鼓励；促进；培养
institute	/'ɪnstɪtjuːt/	*n.* [C]（尤指科学、教育的）机构，研究所
represent	/ˌreprɪ'zent/	*vt.* 代表；提出；表达
recognition	/ˌrekəg'nɪʃn/	*n.* [U] 承认；认可；接受；赏识；表彰
exemplary	/ɪg'zempləri/	*adj.* 典范的；惩戒性的；可仿效的
sustenance	/'sʌstənəns/	*n.* [U] 食物；生计；支持
awareness	/ə'weənəs/	*n.* [U] 意识
socio-economic	/ˌsəʊsiəʊ ˌekə'nɒmɪk/	*adj.* 社会经济的

Background Information

ARCASIA

亚洲建筑师协会（Architects Regional Council Asia，简称亚洲建协或 ARCASIA）是由亚洲最具代表性和最有权威的国家或地区的建筑师学会组成的亚洲建筑师组织。

AAA

亚洲建筑师协会建筑奖（The ARCASIA Awards for Architecture）是亚洲地区建筑界最高建筑设计大奖，与普利兹克奖、金块奖、阿卡汉奖、国际建筑奖、开放建筑大奖并列为面向国际的世界建筑六大奖。

Jakarta

雅加达是印度尼西亚的首都，是世界上人口最多的城市之一。如今的雅加达是印度尼西亚的政治、经济、文化中心以及海陆交通的枢纽，是太平洋与印度洋之间的交通咽喉，也是亚洲通往大洋洲的重要桥梁。

Architectural Terms

region /'riːdʒən/ *n.* [C] a particular area or part of the world, or any of the large official areas into which a country is divided 区域，地区；（国家的）行政区（注：世界或国家的一部分，面积较大，在地理上有天然界限或地域特色的单位，或指行政单位）

- 例句：The *region* produces over 50% of the country's wheat.
  这个地区出产全国 50% 以上的小麦。
- 近义词：

✧ area /'eəriə/ *n.* [C] a particular part of a place, piece of land, or country district 地区；区域（注：普遍意义上的地区，没有明确界限，通常不指行政单位）
词组：mountain area 山区

✧ district /'dɪstrɪkt/ *n.*[C] an area of a country or town that has fixed borders that are used for official purposes, or that has a particular feature that makes it different from surrounding areas 区，区域（注：出于行政、司法、教育等目的而划分的地区）
词组：the business district of Shanghai 上海的商业区

✧ zone /zəʊn/ *n.* [C] an area, especially one that is different from the areas around it because it has different characteristics or is used for different purposes 地带，地区（注：尤指有不同特征或用途的地区，具备独有的特征，有严格的边界）
词组：earthquake zone 地震带

Polysemous Words

单词	常用词义	本文词义
tie	*n.* [C] 领带	*n.* [plural]（ties）纽带；关系，联系

Difficult Sentences

1. The AAA aims at acknowledging exemplary work done by architects working in Asia.

   Sentence analysis

   1) *aim at* 意为"旨在……"

   2) *done by architects working in Asia* 是过去分词作后置定语。

   Translation

   亚洲建筑师协会建筑大奖致力于表彰在亚洲执业的建筑师的杰出成就。

2. It encourages the sustenance of the Asia spirit, the development and improvement of the Asian construction environment and enhancement of the awareness of the role of architecture and architects in the socio-economic and cultural life of Asian countries.

   Sentence analysis

   此句包含三个并列宾语。

   宾语 1：*the sustenance*

   宾语 2：*the development and improvement*

   宾语 3：*enhancement of the awareness*

   Translation

   它（亚洲建筑师协会建筑奖）旨在鼓励亚洲精神的传承，推动亚洲建造环境的发展与改善，增强建筑与建筑师在亚洲各国的社会经济与文化发展中所起的作用。

Text C

Luban Prize

The China Construction Engineering Luban Prize (National Prime-quality Project), hereafter referred to as the Luban Prize, has been established since 1987. It is a selection guided by the Ministry of Housing and **Urban-Rural** Development of the People's Republic of China and implemented by the China Construction Industry Association. The award is the highest honor award for engineering quality in China's construction industry.

The Luban Prize has established high standards for improving the quality of construction projects in China, and put forward an innovative incentive mechanism for industry evaluation of engineering projects. It is of great significance for inheriting the excellent traditions of Chinese architecture, carrying forward the Chinese nation's architectural culture, promoting enterprise technological progress and management innovation, facilitating the upgrading of engineering quality management and improving the core competitiveness of enterprises.

New Words

单词	音标	词义
prime	/praɪm/	*adj.* 首要的，主要的；基本的；一流的
ministry	/'mɪnɪstri/	*n.* [C]（政府的）部
implement	/'ɪmplɪment/	*vt.* 实施，执行
innovative	/'ɪnəveɪtɪv; 'ɪnəvətɪv/	*adj.* 创新的，革新的；新颖的
incentive	/ɪn'sentɪv/	*n.* [C or U] 激励，刺激，鼓励

续表

单词	音标	词义
mechanism	/'mekənɪzəm/	*n.* [C] 机制；原理，途径；机械装置
inherit	/ɪn'herɪt/	*v.* [I or T] 继承
upgrade	/'ʌpgreɪd /	*vt.* 提升，提拔（某人） *n.* [C]（能提高电脑等效能的）升级（或设备）
competitiveness	/kəm'petətɪvnəs/	*n.* [U] 竞争力，竞争性

Background Information

Lu Ban

鲁班（公元前 507 年—公元前 444 年），春秋时期鲁国人，姬姓，公输氏，字依智。木工师傅们用的手工工具，如钻、刨子、铲子、曲尺以及划线用的墨斗，据说都是鲁班发明的。

Ministry of Housing and Urban-Rural Development of the People's Republic of China

中华人民共和国住房和城乡建设部是 2008 年成立的中央部委，是中华人民共和国负责建设行政管理的国务院组成部门，负责国家建设方面的行政管理事务，英文缩写为 MOHURD。

China Construction Industry Association

中国建筑业协会成立于1986年10月，当时名为中国建筑业联合会，英文缩写为 CCIA。

Architectural Terms

1. **urban** /'ɜːbən/ *adj.* of or in a city or town 城市的；城镇的
    - 例句：Today, as *urban* population explodes globally, cities become more crowded.
      今天，随着全球城市人口激增，城市变得更加拥挤。
    - 词组：urban planning 城市规划；城镇规划

urban construction 城市建设；城市工程

urban development 城市发展；城市开发

urban renewal 都市环境改造；都市重建计划

urban jungle 都市生活（尤指城市生活中令人厌恶的部分）

2．**rural** /'rʊərəl/ *adj.* in, of, or like the countryside 乡村的，农村的

- 例句：It was a delightful *rural* scene.
  那是赏心悦目的乡村风光。
- 词组：rural economics 农村经济
  rural electrification 农村电气化
  rural community 农村社会；乡村社区

Difficult Sentences

It is of great significance for inheriting the excellent traditions of Chinese architecture, carrying forward the Chinese nation's architectural culture, promoting enterprise technological progress and management innovation, facilitating the upgrading of engineering quality management and improving the core competitiveness of enterprises.

Sentence analysis

1) *it is of* 后加名词相当于 it is 后加形容词，比如 it is of importance 相当于 it is important。

2) *for* 后连着的 5 个动词 ing 形式（inheriting、carrying forward、promoting、facilitating、improving）作 *for* 的宾语。

Translation

（它）对于继承中国建筑优秀传统、弘扬中华民族建筑文化、推动企业科技进步和管理创新、促进工程质量管理水平升级和提高企业核心竞争力具有重大意义。

Quiz for Chapter 6

Major: ______________ No.: ______________ Name: ______________

1 VOCABULARY AND GRAMMAR

A Check the best meaning for the underlined word in each sentence.

1) ***Laureates*** of the Pritzker Architecture Prize receive a $100,000 grant, a formal citation certificate, and since 1987, a bronze medallion.

A. Winners B. Participators C. Organizers D. Sponsors

2) One objective of the organization is to foster friendly, intellectual, artistic, educational and scientific ***ties***.

A. content B. contract C. contrast D. connection

3) The Luban Prize has been ***established*** since 1987.

A. set out B. set up C. set in D. set off

4) It is a selection guided by the Ministry of Housing and Urban-Rural Development of the People's Republic of China and ***implemented*** by the China Construction Industry Association.

A. implicated B. improved C. executed D. exerted

B Choose the right word.

5) The Pritzker Architecture Prize is an international award ______________ by the Pritzker family through their Hyatt Foundation in 1979.

A. establish B. being established C. established D . be established

6) The purpose of the Pritzker Architecture Prize is to honor a living architect or architects ______________ built work demonstrates a combination of talent, vision, and commitment.

A. who B. whom C. whose D. which

7) I was working as a foreman on a building ______________.

A. position B. site C. location D. address

8) The effects of climate change are likely to be seen across the entire tropical

_______.

A. zone   B. region   C. district   D. area

9) The AAA _____________ acknowledging exemplary work done by architects working in Asia.

A. aims in   B. aims   C. aims with   D. aims at

2 COMMON KNOWLEDGE

10) The Pritzker Architecture Prize is often referred to as "_____________".

A. the Nobel Prize of Architecture

B. the Oscar of Architecture

C. the Grammy Awards of Architecture

D. the Academy Awards of Architecture

11) Jakarta is the capital city of _____________.

A. Indonesia   B. India   C. Indian   D. Indonesian

12) _____________ is the English abbreviation for "亚洲建筑师协会"?

A. ARCASIA   B. ARCA   C. ARCAS   D. ARRCA

13) MOHURD is the English abbreviation for _____________.

A. 中国建筑业协会   B. 中华人民共和国住房和城乡建设部

C. 中国土木工程学会   D. 中国工程师协会

14) _____________ is said to be invented by Lu Ban.

A. The umbrella   B. The saw

C. The boat   D. The broom

15) Which dynasty was Lu Ban in the history?

A. Qin Dynasty.   B. Spring and Autumn Period.

C. Tang Dynasty.   D. Song Dynasty.

16) _____________ often referred to as "the profession's highest honor."

A. The Pritzker Architecture Prize

B. The ARCASIA Awards for Architecture

C. The Luban Prize

D. The Tien-yow Jeme Civil Engineering Prize

Word Power

Major: ________ No.: ________ Name: ________

Write the words and phrases under the pictures.

innovation	architect	rural
bronze medallion	citation certificate	laureate

1. ________

2. ________

3. ________

4. ________

5. ________

6. ________

Group Work

Scan the webpage and complete the chart.

Try to find more information about famous architectural prizes.

Architectural Prizes	Useful Information
The Liang Sicheng Architecture Prize 梁思成建筑奖 THE LIANG SICHENG ARCHITECTURE PRIZE 	Tips: 1. It was established in ______ 2. It is guided by ______ 3. ______ 4. ______
Gold Nugget Award 	Tips: 1. It is considered as the ______ 2. The competition is open to ______ ______ 3. ______ 4. ______
______ **Prize/Award**	Tips: 1. ______ 2. ______ 3. ______ 4. ______
______ **Prize/Award**	Tips: 1. ______ 2. ______ 3. ______ 4. ______

Critical Thinking

In your opinion, what are the key elements to success in your field?

For reference:

Red Boat Spirit (红船精神)

Pioneering: Creation of a New Path and Bold Exploration

首创精神：开天辟地、敢为人先

Persistence: Firm Convictions and Pursuit of the Ideal

奋斗精神：坚定理想、百折不挠

Commitment: Devotion and Loyalty in Service of the People

奉献精神：立党为公、忠诚为民

—— Guangming Daily

Less is more. 以简胜繁。

—— Ludwig Mies van der Rohe

(Golden Prize of AIA in 1960)

Less is a bore. 少即乏味。

—— Robert Venturi

(Winner of Pritzker Architecture Prize in 1991)

Appreciation

Hangzhou Zizhi Tunnel
杭州紫之隧道
the Luban Prize 2017
2020 Tien-yow Jeme Civil Engineering Prize

Hangzhou East Railway Station
杭州东站
2017 ARCASIA awards for Architecture

Hangzhou Olympic Sports Center
杭州奥体中心
2020 Luban Prize

Hangzhou International Expo Center
杭州国际博览中心
2018 Tien-yow Jeme Civil Engineering Prize

Self Assessment for Chapter 6

I can...

Very well	OK	A little	
☐	☐	☐	Use the architectural words in this part
☐	☐	☐	Master some information of architecture prizes
☐	☐	☐	Say something about architecture prizes
☐	☐	☐	Recognize some prize-winning works

参 考 文 献

[1] 林徽因 . 中国建筑常识 [M]. 成都：天地出版社，2019.
[2] 田永复 . 中国古建筑知识手册 [M]. 2 版 . 北京：中国建筑工业出版社，2019.
[3] 张克群 . 中国古建筑小讲 [M]. 北京：化学工业出版社，2020.
[4] 张慈贇，陈洁 . 中国古建筑及其故事 [M]. 上海：上海译文出版社，2017.
[5] XUE C Q, DING G H. Exporting Chinese architecture: history，issues and “One Belt One Road” [M]. Singapore: Springer Nature Singapore Pte Ltd.，2022.
[6] 乐嘉龙 . “一带一路” 上的建筑奇观 [M]. 北京：中国电力出版社，2018.
[7] 李兴钢 . 胜景几何论稿 [M]. 杭州：浙江摄影出版社，2020.
[8] 王澍 . 造房子 [M]. 长沙：湖南美术出版社 ，2016.
[9] 古乔内，哈迪德 [M]. 袁瑞秋，译 . 大连：大连理工大学出版社，2008.
[10] 王拥军 . 董明珠决策格力的 66 金典 [M]. 北京：中国商业出版社，2014.
[11] 杨再德 . 工程建设行业的企业文化管理实践 [M]. 成都：西南财经大学出版社，2021.
[12] 佩尔塔森，翁艳 . 普利兹克建筑奖获奖建筑师的设计心得自述 [M]. 王晨晖，译 . 辽宁：辽宁科学技术出版社，2012.
[13] 中国建筑学会 . 梁思成建筑奖 [M]. 北京：清华大学出版社，2016.
[14] 中国建筑业协会 . 创鲁班奖工程过程精品指南 [M]. 北京：中国建筑工业出版社，2019.
[15] 澳大利亚视觉出版集团 . 扎哈 · 哈迪德和她的建筑 [M]. 付云伍，译 . 广西：广西师范大学出版社，2017.
[16] 故宫博物院古建筑管理部 . 故宫建筑内檐装修 [M]. 北京：紫禁城出版社，2007.